THE UNIVERSE IN A MIRROR

THE UNIVERSE IN A MIRROR

The Saga of the Hubble Telescope and the Visionaries Who Built It

WITH A NEW AFTERWORD BY THE AUTHOR

Robert Zimmerman

PRINCETON UNIVERSITY PRESS
Princeton & Oxford

Published by Princeton University Press, 41 William Street,
Princeton, New Jersey 08540
In the United Kingdom: Princeton University Press, 6 Oxford Street,
Woodstock, Oxfordshire OX20 1TW

press.princeton.edu

Fourth printing, and first paperback printing,
with a new afterword by the author, 2010

Library of Congress Control Number: 2007943159

ISBN: 978-0-691-13297-6 (cloth)
ISBN: 978-0-691-14635-5 (pbk.)

British Library Cataloging-in-Publication Data is available

This book has been composed in Bembo

Printed on acid-free paper. ∞

Printed in the United States of America

10 9 8 7 6 5 4

Oh, I have slipped the surly bonds of earth,
And danced the skies on laughter-silvered wings;
Sunward I've climbed and joined the tumbling mirth
Of sun-split clouds—and done a hundred things
You have not dreamed of—wheeled and soared and swung
High in the sunlit silence. Hov'ing there,
I've chased the shouting wind along and flung
My eager craft through footless halls of air.
Up, up the long, delirious, burning blue
I've topped the wind-swept heights with easy grace,
Where never lark, or even eagle flew;
The high, untrespassed sanctity of space,
Put out my hand, and touched the face of God.

John Gillespie Magee, Jr.

CONTENTS

ILLUSTRATIONS

COLOR SECTION

Supernovae

Eta Carinae

PREFACE

It has without question been the grandest instrument that humans have ever sent into space. For more than a decade the Hubble Space Telescope has churned out image after image, each fundamentally changing the public's perception of the universe in unexpected ways.

Conceived in the 1940s and 1950s, its gestation was long and difficult, blocked by naysayers and doubtful scientists who feared its cost.

Designed in the 1960s, its birth was long and difficult as engineers, astronomers, and bureaucrats fought over its design.

Built in the 1970s and 1980s, its childhood was at first crippled, as a fundamental error in construction left its mirror defective.

Fixed and maintained in the 1990s by high-flying astronauts who loved it as much as if not more than the scientists who used it, Hubble lifted a curtain from our view of the universe, changing it so profoundly that no human can look at the stars in the same way again. At the same time Hubble unexpectedly transformed the manner in which both astronomers and astronauts do their work.

None of this would have happened without the unceasing dedication of a host of individuals, most of whom are and will unfortunately always remain nameless. Many sacrificed years to the telescope. Some even ruined their lives and careers to get it built. A few did everything possible to get it launched, only to be left on the wayside when the telescope was finally in space and working.

This book is my attempt to tell their story, to make known a few of the men and women who conceived, designed, built, repaired, and saved

Hubble over the decades. In doing so I found myself telling a story of how human beings can sometimes be shortsighted and foolish, and how they can more often rise above that foolishness to make great things happen.

I tell this story not as an astronomer, which I decidedly am not, but from the perspective of a science writer and space historian who has viewed Hubble's output with the same amazement experienced by most ordinary people. This perspective is important, since it is ordinary people who have paid for this optical space telescope and want its capability maintained. Astronomers, who often have very different reasons for building telescopes and sometimes justifiably do not consider an optical telescope their most important tool, would be well advised to listen to this perspective if they want to keep their government-funded science budgets healthy and growing.

I also tell this story with a view toward the future. In the coming decades the human race will take the first tentative steps toward establishing permanent colonies in space. When that happens, the stars above will beckon in ways that we on Earth cannot yet imagine. Living on the Moon, or in a space station, spacefarers will see the sky in all its glory, whenever they look up. It will become an essential and dominant part of their normal landscape, its splendor omnipresent and unavoidable. For pleasure, personal fulfillment, or scientific research, these spacefarers will build all kinds of telescopes, aiming them skyward to blink in wonder at the Trapezium in Orion, the Great Andromeda Galaxy in Andromeda, the Ring Nebula in Lyra, and the ever changing weather on Jupiter and Saturn.

When that happens, the human perception of the universe will undergo as fundamental a change as Galileo experienced when he first aimed a telescope at the stars. No longer will our vision of the heavens be limited to a single optical telescope orbiting the Earth. For the first time, we will have many eyes peering directly out into the unknown blackness above, and for the first time, we will truly begin to perceive the Earth's place in the cosmos.

The Hubble Space Telescope gave us our first hint of what that existence will be like. We should not, therefore, forget the effort of those who made that hint possible.

■ ■ ■

This book would not have been possible without the generous help of the astronomers and engineers who built and continue to use and operate Hubble. Special thanks must be extended to Bob O'Dell, Sandy Faber, Jeff Hester, Ray Villard, John Wood, and everyone at the Space Telescope Science Institute as well as numerous astronomers everywhere, whose never-ending willingness to answer my endless questions always amazed me. Thank you all, again.

I must also thank the many dedicated librarians and archivists who work ceaselessly and without fanfare to keep alive the past. Specific thanks must go to Robin Dixon of the Homer Newell Library at the Goddard Space Flight Center, Marilyn Graskowiak and Mark Kahn of the National Air and Space Museum Archives, Steve Dick, Jane Odom, John Hargenrader, and Colin Fries of the NASA History Office in Washington, and James Stimpert at the Milton S. Eisenhower Library at Johns Hopkins University. Without their generous assistance I could not have written this book.

Thanks must also go to my editors, Jeff Robbins and Ingrid Gnerlich, who were both willing to say yes to the idea.

And I mustn't forget to say thank you to my wife Diane, who knows when to leave me alone when I need to work.

Finally, I must acknowledge the men and women who have been willing to risk their lives to fly into space and fix Hubble. Without their effort, none of Hubble's discoveries would have happened and the horizons of the human race would surely be more limited. It is they, as much as anyone, who make our future possible.

ROBERT ZIMMERMAN
Beltsville, Maryland

THE UNIVERSE IN A MIRROR

1

Foggy Vision

The sky was dark, the air clear. It was an excellent night for astronomical photography.

On March 7, 1945, Enrique Gaviola of the Cordoba Observatory of Cordoba, Argentina, carefully positioned the observatory's 61-inch telescope for an evening of research. Painstakingly, methodically, Gaviola aimed the telescope at one of the more spectacular spots in the southern sky, the Keyhole Nebula in the constellation Carina.

First observed by John Herschel in the mid-1830s while in South Africa doing a survey of the southern sky, it had been given its name by Herschel because of its distinctive keyhole-shaped dark patch. What made this particular place in the sky even more intriguing was that on December 16, 1837, Herschel had been surprised to see a new star shining brightly there. "[The star] had come on suddenly," he wrote that night in a letter to Thomas Maclear, the astronomer at the Royal Observatory at Cape Town.

At first Herschel thought the star might be what was then called a bright nova, similar to those discovered in 1572 and 1604, and now dubbed supernovae. After some careful measurement, however, he realized that the gleaming, unexpected spark above him was not a new star suddenly bursting into visibility, but the star Eta Carinae, shining three times brighter than he had ever seen it before, and approaching 1st magnitude.

For several hours into the wee hours of the morning Herschel stared at this inexplicable object. For years it had remained unchanged to him, shimmering at about 2nd magnitude just off the edge of the Keyhole itself. In fact, only a week earlier he had noted the star's annual arrival in the December evening sky, and had commented to his chief assistant that "We must soon begin [studying] him again."

Before Herschel could "begin," however, the star had suddenly become one of the brightest in the sky. Unable to contain his excitement, he called his wife, his assistant, and his personal butler all out of bed to have them look and confirm what he saw.

As he wrote that night to Maclear, "How big *will* it grow?"

In the thirty years that followed Eta Carinae faded in fits and starts from 1st to 7th magnitude, while the darker parts of the much larger Keyhole Nebula slowly brightened so that it no longer stood out so distinctly.[1]

In the twentieth century astronomers returned to this star periodically, trying to figure out what had happened in 1837 as well as afterward. Though some thought the nebulosity surrounding the star was slowly growing, in 1932 astronomer Bart Bok of Harvard concluded decisively that this was an imaginary effect.[2] In fact, most observers in the early twentieth century assumed that the faint collection of bright hazy spots surrounding Eta Carinae were actually a handful of individual stars, embedded in a gas cloud.

Now, more than a hundred years after Herschel, Enrique Gaviola was back, taking another careful look at Eta Carinae. For several hours he took two sets of nine images, beginning each set with a one-second exposure and doubling the exposure duration each time until the exposure for his last picture was over four minutes long.

Once developed, these images from 1945 were considered by many the best ever taken of this strange star. They showed what Gaviola humorously dubbed the Homunculus, Latin for "little man," a kind of Pillsbury Doughboy "with its head pointing northwest, legs opposite and arms folded over a fat body."[3]

As good as Gaviola's photographs were, they were generally fuzzy and revealed little detail. The best that Gaviola's images could do was to show that the hazy bright spots surrounding Eta Carinae were probably not multiple stars but several gas shells and clouds enveloping the star and illuminated by it. Moreover, he was able to extrapolate backward and conclude that the nebula was formed "by clouds ejected by the star around 1843," about the time of a second outburst following Herschel's initial 1837 discovery.

Why the expansion happened, how it was unfolding in detail, and where it was going to end up, however, was utterly impossible for Gaviola to deduce from his nebulous images. Gaviola's photographs, as

Fig. 1.1. A sequence of four images of Eta Carinae, taken by Enrique Gaviola, March 7, 1945. The exposure times, from left to right, are 256, 128, 64, and 32 seconds. In all the Eta Carinae images, north is up and east is to the left. (Photo by Enrique Gaviola, provided courtesy of Arnout van Genderen)

groundbreaking as they were, were typical of all astronomical images since the invention of the camera. The atmosphere that we breathe and that makes life possible also acts as an annoying translucent curtain, blurring our vision of the sky. Just as a prism will bend the light that passes through it, so does the atmosphere. The atmosphere, however, is in constant flux, causing the path of that light to shift and jiggle. When we look up at a star, this shifting makes it appear to twinkle. On a photographic plate this twinkling in turn causes the accumulated light to spread out so that a truly sharp image is just not possible.

The result: before the advent of space flight astronomers, both professional and amateur, were left thwarted and unsure about what they saw. For someone like myself, who has poor vision and requires glasses, this situation is self-evident. Though the metaphor is not technically correct, for me to understand the limited view of the heavens from the beneath the atmosphere, all I have to do it is to take off my glasses. Everything becomes fuzzy, unclear, and indistinct.

I, however, can buy eyeglasses. Until the late twentieth-century astronomers had no such option. Trapped on the Earth within its unsteady and hazy atmosphere, astronomers were condemned to look at the heavens as though they had bad vision and were forbidden from using glasses.

The consequences of this hazy situation have been both frustrating and profound. Consider for example the efforts of Giovanni Schiaparelli and Percival Lovell to map the surface of Mars in the late nineteenth and early twentieth centuries. Beginning in 1877 Schiaparelli studied Mars nightly, using an 8.6-inch telescope at the Brera Observatory in Milan. After more than a decade of work he finally published his map, outlining a wide range of vague shapes and streaks on the Martian surface. Of greatest interest were what he called "*canali*" (which means "grooves" in Italian).

Though Schiaparelli was convinced the *canali* were real, he found that their

> . . . aspect is very variable. . . . Their appearance and their degree of visibility vary greatly, for all of them, from one opposition* to another, and even from one week to another . . . . often one or more become indistinct, or even wholly invisible, whilst others in their vicinity increase to the point of becoming conspicuous even in telescopes of moderate power.[4]

Percival Lowell followed Schiaparelli with decades of more work, studying Mars's surface and making endless sketches of what he thought he saw there.

From Lowell's perspective, the complex series of straight lines that crisscrossed Mars strongly suggested what could only be artificial constructs, which he labeled more bluntly as canals. As he wrote in 1895, "There is an apparent dearth of water upon the planet's surface, and therefore, if beings of sufficient intelligence inhabited it, they would have to resort to irrigation to support life."[5] To Lowell, the canals appeared to be built by the inhabitants of the red planet as a vast irrigation system to stave off the consequences of an increasingly arid planet.

For the next seventy years the human race debated the possibility of life on Mars. Lowell's thoughts inspired such classic works of fiction as H. G. Wells's *War of the Worlds* as well as a plethora of science fiction books and movies.

Then, in the mid-1960s the United States sent a series of unmanned probes to Mars to take the first close-up images—and burst Lowell's bubble. Mars has no canals, no intelligent life. The canals were an optical illusion created by the Earth's varying atmosphere.

The atmosphere causes similar problems across the entire field of astronomical research. Worse, not only does it distort optical light, it en-

* Opposition is the moment each year when the Earth is exactly between the Sun and Mars, and therefore best positioned for viewing.

tirely blocks large portions of the rest of the electromagnetic spectrum. Except for radio wavelengths and a few select infrared wavelengths, the majority of the infrared and all of the ultraviolet, x-ray, and gamma ray regions of the spectrum are inaccessible to astronomers working from the surface of the Earth. This fact is especially crippling because the bulk of astronomical research is done through spectroscopy, and much of the most interesting and informative spectroscopy needs to be done in these unavailable wavelengths.

For example, by observing the spectrum of light coming from a star, astronomers can gather information about that star's chemical makeup. Each element when heated emits light at a specific wavelength. If you see a spike of light at that specific wavelength you know that element is present in a star's atmosphere. Similarly, if there is a dip of light at that wavelength you know that that element is standing somewhere between you and the star, either in the star's surrounding nebula or in some intervening gas in interstellar space, absorbing that light. Unfortunately, the spectral signature of a large percentage of the most interesting elements occurs at wavelengths outside visible light in parts of the electromagnetic spectrum that are blocked by the atmosphere.

In the visual wavelengths, meanwhile, the atmosphere's blurring action makes the interpretation of the astronomical data more challenging. Our brains are tightly wired to our eyesight. Very roughly speaking, if we cannot see a clear image of something, it is difficult for us to fully grasp what is going on, no matter how much other information is available. Conversely, if we have a good image to look at, we can more easily interpret all the other data and understand how they fit into that visual image.

Consider for example what astronomers call planetary nebulae. These objects were given that name because at first glance they seemed to resemble planets, but by the 1800s scientists had realized that the nebulae were not planets at all but distant stars surrounded by large and beautiful cloud structures.

By the early 1960s astronomers were reasonably sure they understood their origin. When a star like the Sun has used up most of its hydrogen fuel and begins burning helium, it becomes unstable, starts to pulse, and ejects mass in a series of expanding shells. After some ten to fifty thousand years these shells form a planetary nebula, which surrounds a slowly dying and cooling white dwarf star.

This theory, however, did little to explain the complex but hard-to-see structure of the encircling clouds visible in pre-Hubble astronomical photographs. In most cases the nebula looked like one or several rings. For example, in long photographic exposures the Ring Nebula in the constellation Lyra resembled a bluish-green oval with horizontal wreathlike veils cutting across its central regions. Similarly, the Helical Nebula in Aquarius looked like two overlapping rings, though the best photographs also showed strange spokelike features pointing inward toward the central star. Other planetary nebulae, such as the Dumbbell Nebula in the constellation Vulpecula, looked as if we were viewing the ring edge-on so that it resembled a barrel on its side.

Because so many of the nebulae had this ringlike morphology, it was assumed that they were really shells or bubbles, with only the outer edges visible because our line of sight was looking through the most material. Such an assumption conformed nicely to the idea that the shells were the debris from the star's earlier helium-burning stage, when it repeatedly ejected large amounts of mass.

Other planetary nebulae, however, did not conform to this theory. Some appeared irregular, patchy disks with no discernible pattern. Others had weird shapes, making any interpretations difficult if not impossible. For example, the Owl Nebula in Ursa Major had an outer ring, but instead of an open interior its central regions looked more like an hourglass, two conelike shapes pointing inward toward the central star. And the Saturn Nebula in Aquarius was even more baffling: it had two rings, each inclined at a different angle to our view. Especially baffling were the two spikes of material at opposite ends of the nebula pointing away from the central star. For these inexplicably shaped planetary nebulae, several theories were proposed to explain their formation, including the possibility that the spikes were jets emanating from the poles, or some form of slow expansion influenced by either magnetic fields or unseen binary companions.

Because the images were so fuzzy and indistinct, however, it was difficult for scientists to reach a consensus on any specific theory. And though spectroscopy provided a great deal of information about the motion within each nebula's surrounding gas cloud, it was often difficult to untangle this spectral velocity data into a coherent picture without a corresponding sharp visual image. Thus, few astronomers made a serious effort to explain the formation of these nebula shapes because the data were so imprecise.

The problem was the same for galactic evolutionary theories. It was impossible with ground-based telescopes to see any galaxies from the early universe, and thus get a longer view of the evolution of galaxies across time. These distant objects were simply too faint to be picked out from the blurring effects of the atmosphere. Similarly, there were a number of very strange-looking distant galaxies, such as the Antennae Galaxy in the constellation Covus, with its two long trailing tails and two warped central blobs, whose shapes ground-based telescopes could not image sharply. Though astronomers were able to put together a number of theories about galaxy mergers or collisions to explain these unusual structures, any one of these ideas could be right. Worse, until better and more precise data were available, including information from the wavelengths blocked by the atmosphere, it was also quite possible that none were correct.[6]

In various areas of astronomy this problem repeated itself. Astronomers could put together reasonable theories to explain their data, but without clear optical images it was difficult to confirm which theories were the most accurate.

For the general public, the situation was worse. Dependent as we humans are on our eyesight, the atmosphere essentially left the human race blind to the heavens. We were like a nearsighted man before the invention of eyeglasses. We could squint and strain and maybe make a guess at what we were looking at, but to actually perceive the reality of the universe in all its glory was nigh on impossible.

■ ■ ■

Even as Gaviola was slowly developing his photographs and preparing his paper for publication—crippled as he was by being at the bottom of a 100-mile-thick fog filter—another astronomer almost half a world away was about to take the first step in what would become an epic, half-century-long odyssey to solve this centuries-long dilemma. This man was about to propose that the United States build the first optical telescope in space.

World War II had just ended. At the time Lyman Spitzer, Jr., was a thirty-one-year-old astronomer doing war work as head of a research organization called the Sonar Analysis Group. Though most of his group worked in the Empire State Building in New York, Spitzer's headquarters and base of operations was in Washington, DC. As Spitzer explained in a 1978 interview, "My work involved talking with people who were doing [sonar] research and telling them what they were doing wrong and what they ought to be doing."[7]

Before the war Spitzer had been a young post-graduate astronomer working at Yale University. Now that the war had ended he wanted to get back to astronomy work.

In the fall of 1945, however, Spitzer was still working in Washington, DC. Among the many scientists he ran into in DC who were part of the war effort was a geophysicist named David Griggs. During the war Griggs had been part of a group of scientific advisors working under Dr. Edward Bowles, who had been named special assistant to Secretary of War Henry Stimson. Under Bowles's leadership, Griggs and his cohorts had been key on-site technical advisors during the D-Day invasion, the campaign in France, and later during the Battle of the Bulge. As noted by historian James Baxter, "They evacuated equipment at the last moment, they served as pinch-hit operators of gear in crucial spots, often under fire."

Now that the war was over, the military research group that Griggs was involved with was undergoing a reorganization. He explained to Spitzer how the Air Force was forming a new secret group at the Douglas Aircraft factory out in Santa Monica, California, called the RAND Project (for Research ANd Development). Though not yet finalized, RAND's first report for the Air Force was to be on the benefits of rockets and orbiting satellites and titled "Preliminary Design of an Experimental World-Circling Spaceship." Griggs, active in the development of this project, asked Spitzer if such a spaceship could have uses for astronomy.[8]

Spitzer was immediately intrigued. To him, the idea of putting a telescope in space was both scientifically and emotionally appealing.

Over the next few months, as he wrapped up his war work and returned to teaching astronomy at Yale University in New Haven, he kept in touch with Griggs and others in Santa Monica, letting them know that he was interested in providing his input should the project get started. When on March 2, 1946, the Air Force and the Douglas

Fig. 1.2. Lyman Spitzer with his children, Sarah, 1, Dionis, 5, and Nicholas, 8, on Nicholas's birthday in Princeton, 1950. (Photo courtesy of Sarah Lutie Spitzer Saul)

Aircraft Company signed a $10 million contract to form RAND, Spitzer quickly made arrangements to spend a week at the Project RAND headquarters in the Douglas Aircraft factory, where he wrote a paper for the project called "Astronomical Advantages of an Extra-Terrestrial Observatory," describing in detail the scientific advantages of building a telescope in space.

Spitzer was by far not the first to suggest the advantages of placing a telescope above the Earth's atmosphere. Hermann Oberth of Germany was the first to describe the advantages of building a telescope in space in his 1923 groundbreaking book, *Die Rakete zu den Planetenraumen* ("By Rocket into Planetary Space"), originally written as his doctoral dissertation but rejected by his school advisors and then published privately. In 1933, Henry Norris Russell, the director of the Princeton University Observatory and the man under whom Spitzer had gotten his degree, bemoaned his inability to do ultraviolet spectroscopy because of the Earth's atmosphere, and dreamed of an astronomer's heaven where he was "permitted to go, when he died, instruments and all, [to] set up an observatory on the Moon." Then in 1940, writing for the science fiction magazine *Astounding Science Fiction*, astronomer Richard Richardson proposed his own concept for building of a 300-inch lunar telescope.[9]

What made Lyman Spitzer's 1946 paper different, however, was that it was concrete, realistic, and based on technology that was either available at the time or expected to be developed in the coming decade. He was not speculating or exercising a mere flight of fancy. He was applying the increasingly available new technology of rockets—demonstrated by the V2 rocket during the war—and suggesting it be used to place a telescope in space.

Nonetheless, Spitzer's proposal was hardly conservative. Though he described the possibilities of research using an orbiting 10-inch telescope, he quickly went on to propose the construction of something that was far more ambitious, an orbiting reflecting telescope with a mirror 200 to 600 inches in diameter.

You have to understand the context of this proposal to realize how audacious it was. In 1946 the 200-inch Hale Telescope on Palomar Mountain in California, soon to become the largest ground-based telescope in the world, was not yet finished. It had taken almost three decades to build, and it would not even be dedicated until a year later. Moreover, when finished it would weigh a million pounds and be almost seventy feet tall. In addition, in 1946 when Spitzer wrote his report, the first orbiting satellite was still more than a decade away, and that spacecraft—Sputnik—would weigh a mere 185 pounds.

Yet here was Lyman Spitzer proposing that the United States not only consider building a telescope as much as three times bigger than the Hale

Telescope *but also put it in orbit around the Earth.* As Spitzer noted in his report, such a project would not only provide humanity with its first clear view of the heavens, it would more importantly "uncover new phenomena not yet imagined, and perhaps . . . modify profoundly our basic concepts of space and time."[10]

At first glance Lyman Spitzer did not impress people as being such a wild-eyed dreamer. Tall, thin, and gangly, his soft-spoken and gentle manner gave one the impression that he was happier buried among a pile of books than pushing the risky unknown. Moreover, he had spent almost his entire life in the academic world.

Spitzer came from traditional New England stock, his ancestors first arriving in America in the mid-1700s. His father had gone to school at Andover, then Yale, then became a successful and wealthy businessman, first as a municipal bond salesman and then as the owner of a paper box factory in Toledo, Ohio.

With that money A. Lyman Spitzer, Sr., was able to travel, taking his family on trips to France, Switzerland, England, California. In 1925–26, when Spitzer junior was eleven, the family lived in Paris for six months. Later they spent four months in Rome. "We got around a bit," Spitzer remembered in 1977.[11]

Following in his father's footsteps, Spitzer started his studies at Andover and continued at Yale. After this, however, Spitzer didn't go into business like his father, but continued in academia, going to Cambridge University in England on a scholarship, then Princeton, where he earned his PhD, then Harvard as a fellow, and then back to Yale as a teacher. By the time he was thirty-three he was the chairman of the Princeton astrophysical sciences department, taking over for Henry Norris Russell, his academic mentor.

As privileged as Spitzer's upbringing might seem, he did not grow up spoiled. For Spitzer, astronomy and intellectual studies were a natural passion, and he pursued them relentlessly. Moreover, he had an ardent fascination with the idea of doing things that no one had ever done before. While in college he became fascinated with science fiction, and dreamed up his own transcontinental transportation system using electromagnetic suspension. "Small cars would travel in tubes between cities, and end up various places within the city, and might even, in tall buildings, go up and stop at one of the high floors," he explained in 1977.

Yet, even as he fantasized about building this vast transportation network, Spitzer also recognized, with an easygoing and witty self-depreciation that made people like him, how wild-eyed the fantasy was. "I told my father about this, and he began to think I was going off the deep end."[12]

Through it all, Spitzer always seemed to keep a placid and good-natured view of the world. "I've never been a fighter, by profession," he mused in 1977. "I go out of my way to keep things on a friendly basis . . . . You can have controversy without being unfriendly."[13]

Despite Spitzer's upscale and bookish background, he was a remarkably fearless and athletic man. For example, on July 28, 1945, he was working in an office on the 64th floor of the Empire State Building—he liked to joke how he hunted enemy submarines from these heights—when an Army B-25 bomber got lost in the fog and crashed into the north side of the building, plowing into the 78th and 79th floors.

Because the windows had been closed Spitzer only heard the zoom of the plane, which for some strange reason got cut off suddenly. Then, even more puzzling, he could see debris falling past his windows. With almost childlike curiosity, Spitzer walked over to the windows and started to open one, intent on peering out and up to see what had happened.

Another scientist, Peter Bergmann, had to actually hold him back, convincing him that this was not a good idea. "He was, of course, perfectly right," Spitzer admitted cheerfully in 1978.[14]

His athletic skills became more evident after the war, when his love of the outdoors got him interested in the hobby of mountain climbing. At first he and his wife Doreen would take hiking trips to Europe and the Alps, exploring the mountainous regions while Spitzer looked longingly at their peaks. Then, in 1955 they arranged a guided trip to the Alps. After climbing a series of increasingly challenging mountains they capped their adventure with an ascent of the Matterhorn.

Once back in the States Spitzer began making regular caving and rock-climbing trips with his graduate student Don Morton. One time, in a letter to his family describing a recent very challenging mountain expedition, Spitzer wrote, "You may wonder what I find enjoyable in a mountaineering trip of this sort, and I confess I find myself asking this same question. Certainly most of the trip was not particularly comfortable. . . . Much of the time I was looking forward to the end of whatever I was doing. Yet I find a certain satisfaction in undertaking an adventure of this sort, and in pushing myself to the maximum effort."

"He loved it," Doreen Spitzer remembered. "It was a very great relief to him. . . . The challenge took his mind off of what he was doing." Spitzer's daughter Lutie Spitzer Saul explained her father's passion for mountains and rock climbing in another way. "For some people these heights are a substitute for spirituality."[15]

With such a bold personality, it is perhaps not surprising that Spitzer was willing in 1946 to propose building a telescope in orbit that was two or three times bigger than anything that had yet been built on Earth.

Spitzer's proposals were too farsighted, however, to gain acceptance, despite what seemed an enthusiastic response within government circles to this first RAND Corporation report. Throughout the late 1940s and most of the 1950s Spitzer found little interest in his space telescope idea. During those years before Sputnik, he spent most of his research time studying the empty regions of space between the stars and galaxies—trying to figure out the nature and makeup of these almost empty clouds of dust and gas from which new stars were thought to form—or building one of the first attempts to create a controlled fusion reactor, something he called the Stellarator.

This second classified project, dubbed Project Matterhorn in honor of his Matterhorn climb, was as farsighted as anything else Lyman Spitzer ever proposed. The idea was to build the first "magnetic bottle," designed to contain a gas made up of deuterium at 100 million degrees Kelvin long enough for a controlled nuclear fusion reaction to occur. As he wrote forty years later in a New York Times op-ed, "If we could replicate the process that powers the sun, we could create a source of virtually unlimited energy."[16]

The Stellarator was something right out of a 1950s science fiction movie. A tube two to four inches wide and ten to twenty feet long was twisted into an endless figure-eight shape and then charged with gigantic amounts of electrical energy. "Since the power required at the peak of the field is in the neighborhood of 50,000 kilowatts," Spitzer wrote in 1958, "the power bill has restricted operations to pulses lasting about 0.02 second."[17]

Fig. 1.3. Lyman Spitzer rock climbing in the Shawangunk Mountains, New York. (Photo by Don Morton)

In between building several Stellarators and his interstellar research, however, Spitzer never abandoned the idea of space exploration and its uses for astronomy. Periodically he would write carefully thought out papers for the journals of such organizations as the American Rocket Society or the British Interplanetary Society, describing the construction of a nuclear ion engine for traveling between planets or working out the orbital mechanics of a small satellite in a circular orbit around the Earth. Other times he would appear at conferences, advocating the idea of space exploration and its advantages.[18]

Though few people expressed strong hostility to his ideas, few showed much support, either. Scientists were especially skeptical. After one of his conference presentations a scientist came up to him and said, "Lyman, I admire your courage." Though he liked what Spitzer had said, he considered it somewhat far-fetched. "Most astronomers didn't take it seriously," Spitzer remembered in 1977. "They thought I was sort of . . . wild-eyed or wide-eyed, one or the other."[19]

Some astronomers were more harsh. In 1953, when astronomer Gerald Kuiper heard of Spitzer's proposals for space-based astronomy, he said, "I would regard the [funding] of this project hazardous and probably undesirable."[20]

Still, Spitzer persevered, often inspiring others into action and getting them to do things they would never have imagined doing. For example, one day in 1954, Spitzer was having lunch with two fellow scientists, Martin Schwarzschild and James Van Allen. Schwarzschild was a fellow professor in Princeton's astrophysics department, which Spitzer headed. Van Allen in turn was at the time one of the country's most respected space scientists, having used the V2 rocket extensively in the postwar era to do the first studies of the Earth's magnetosphere.

Van Allen was then on a temporary sabbatical from the University of Iowa to work with Spitzer on Project Matterhorn. Because the project was classified, Spitzer had been having trouble hiring good people. "It was difficult to add staff in those days because we couldn't say what we were doing, and our salary scale wasn't that high." Putting the very well known Van Allen in charge of the experimental group made it easier to convince others to join.[21]

Schwarzschild meanwhile had just returned from a sabbatical doing astronomical observations at the Mount Wilson Observatory. The son of the German physicist Karl Schwarzschild (who is most famous for taking Einstein's equations and using them to describe the environment around a black hole), Schwarzschild had fled Germany in 1936 because as a Jew he had been banned from working at any German university.

He came to the United States with fear and trepidation. "I did not want to spend my life [there]. . . . I had a simplified picture, to exaggerate a little, that the United States consisted of Indians, gangsters, and Mount Wilson."[22]

In the end he grew to love America more than many of its natives did. After Pearl Harbor, he immediately enlisted in the Army, going in as a buck private and ending up as an officer on special assignment with the Air Force on the front in Italy, where he analyzed the effectiveness of U.S. bombing. Though assigned a New York truck driver to get him around, his German accent more than a few times got him arrested as a German spy. As astronomer Virginia Trimble noted in her 1997 obituary of Schwarzschild:

> One can imagine the reaction of the officers *in situ* when asked by a stranger with a heavy German accent, "Please tell me how your bombs are aimed," and he spent an occasional night in the brig, maintaining his usual cheerful calm, partly to avoid embarrassing his captors when the truth came out. Sorting things out at various times involved checks with headquarters, the intervention of an English officer on similar assignment, and a New York truck driver, whose primary assignment seems to have been to say, slowly and firmly in suitable dialect, "Ee's OK, see."[23]

Upon returning home Schwarzschild received many university job offers. "Very flattering but also very complicated to decide," he remembered in 1977. Rather than take a job as a department head ("I didn't trust that I had the judgment"), he decided instead to "go to the place with the best head."[24]

Meanwhile, Spitzer was being considered for the job of running Princeton's astrophysical department. Harlow Shapley, director of the Harvard College Observatory, had been acting as a mediator between Spitzer and Princeton. During negotiations Shapley asked Spitzer to outline in detail the conditions under which he would seriously consider coming to Princeton. In answer, Spitzer put together a long-range plan describing his intentions for the department and sent it to Shapley. Included in that plan was his desire to hire Martin Schwarzschild as a full professor. "I'd always been a great admirer of Martin Schwarzschild's since I first met him," Spitzer noted many years later. "He always seemed such an incisive, enthusiastic, clearly organized scientist."[25]

For Schwarzschild the feeling was mutual. "I wanted to be in a department led by Spitzer."[26] In 1947 he joined the Princeton astrophysical department under Lyman Spitzer's leadership.

Also part of Spitzer's master plan was his insistence that he and Schwarzschild alternately spend one semester every two years away from the university doing observational research. For years afterward they would each spend half a year at the Mt. Wilson observatory in Pasadena, California, using its 100-inch Hooker telescope.[27]

The 1954 lunch with Schwarzschild, Spitzer, and Van Allen took place immediately after Schwarzschild's most recent trip out west, where he had been working with, of all people, Richard Richardson, the astronomer and sometime science fiction author who in 1940 had written an article proposing the construction of a 300-inch telescope on the Moon for the science fiction magazine *Astounding Science Fiction*. The two men had been trying to photograph the convective turbulence on the surface of the Sun. Both of them had been very frustrated, however. As Schwarzschild explained during that lunch with Spitzer and Van Allen, "I complained bitterly about the hard fate of the astronomer sitting under this miserable atmosphere."

Schwarzschild remembered Van Allen laughing and saying, "Oh, you astronomers should just get off your traditional ways and send your telescopes up in balloons. We cosmic ray physicists have done it for a decade or two, with quite complicated instruments. You are just too ground-bound."[28]

The idea of using telescopes on balloons, which strangely enough had not occurred to either Spitzer or Schwarzschild, excited them both. Spitzer unfortunately couldn't spare the time for such a project, committed as he was to the Matterhorn project.

Schwarzschild in turn was not an experimentalist, and was doubtful he could do it. Over the next few months Spitzer pressed him, however. "Why don't you try?" he would say in his gentle but insistent manner.

Schwarzschild could not resist, and with Spitzer's help he spent the next four years building the first balloon-borne telescope, called Stratoscope.* While Schwarzschild ran the project, Spitzer did the fund-raising, getting the Office of Naval Research to finance the project. For the construction of the telescope Schwarzschild contracted a Connecticut company called Perkin-Elmer, known for building high-precision scientific and military optics. For its guidance system he hired what became Ball Brothers, later known for building some of NASA's best scientific and military satellites. For the balloon, he hired a balloon company in Minneapolis. "The whole setup," Schwarzschild remembered, "when you look from the present point of view, was fantastically primitive."[29]

Nonetheless, in the summer and fall of 1957, Schwarzschild's 12-inch balloon telescope made a handful of flights, taking tens of thousands of pictures. The first flight, on August 22, 1957, had a "hair-raising launch" according to Spitzer, though it successfully carried a dummy telescope to test the guidance system. The second flight, on September 25, took the balloon to 80,000 feet, where it took some 8,000 pictures of the Sun. A third flight in October was reconfigured to produce five slow scans of the Sun's surface.

Schwarzschild's results were mixed, but nonetheless exhilarating. "Mostly with nothing, but a few frames of entirely superior quality," Schwarzschild remembered. "[They] were the first off-the-ground astronomical [images]." The pictures showed for the first time the polygonal granulations that churn about on the Sun's surface. As noted by *Sky and Telescope*, this success "foreshadow[ed] many kinds of future observations in which the astronomer is on the ground while his equipment is taken above the atmosphere to where observing conditions are nearly perfect."[30]

Schwarzschild's project was part of what was to be one of the most important international scientific endeavors ever attempted, called the International Geophysical Year. Organized by scientists in the mid-1950s, the IGY intended to encourage researchers worldwide to simultaneously study "the fluid envelope of our planet—the atmosphere and oceans—over all of the Earth and at all heights and depths." The IGY's time period,

* Ironically, Schwarzschild found out years later that his own father had attempted to do the same thing in Germany, using a Zeppelin.

from July 1957 through December 1958, was picked to correspond to the solar maximum in the Sun's eleven-year solar cycle, the period when the Sun's sunspot activity is at its most intense. As part of the event, the organizers not only called for global studies of the Sun, the weather, the Earth's magnetism, its aurora, and its geology, they also issued a challenge to the participants to build and launch the first artificial satellite.

As obvious and as enthusiastic as many of the United States' scientists were about the idea of orbiting a satellite, there was also a great deal of skepticism. Some worried that the general public would not understand the event and would somehow see it as dangerous. Others fretted about the cost, which was certainly several magnitudes greater than what the United States was spending to launch suborbital sounding rockets to do atmospheric research.

After several months of debate, scientists from a number of committees at both the National Research Council and the National Science Foundation eventually agreed to work together to convince the U.S. government to build a satellite. This decision was then followed by more negotiations, this time between these quasi-governmental academic organizations and the federal government. Finally, in late July 1955 the White House announced that the United States would launch a very small satellite, called Vanguard, as part of the IGY.[31]

From Spitzer's point of view, Vanguard was certainly a step in the right direction in his dream of building a space-based telescope. Nonetheless, Vanguard was very small, a sphere less than seven inches in diameter and weighing slightly more than three pounds. Moreover, the United States had no clear plans to do anything in space beyond Vanguard. Considering the skepticism that Spitzer had seen from astronomers about his ideas, it didn't appear that the construction of an orbiting space telescope would occur anytime in the near future.

Then, on October 4, 1957, everything changed. On that day, the Soviet Union—not the United States—proved to the world that a space satellite could be built. And they did it with a satellite that weighed seventy times more than Vanguard and was three and a half times bigger.

On that day they launched Sputnik.

2

Slow Start

Geoffrey Burbidge was puzzled. It was 1961 and for the last four years, he and his wife Margaret had been astronomers at the Yerkes Observatory. During the winter months Yerkes was a very isolated place, located as it was 80 miles north of Chicago, Illinois, just north the border into wintry Wisconsin near the small town of Williams Bay.

The one thing that Yerkes was blessed with, however, was a great library. During the cold fall and winter months of 1961 Burbidge spent a lot of time browsing through old and new journals, not just to learn what had been studied in the past but to keep up with the research being done at the moment.

One 1956 paper especially caught his eye. An astronomer at the Radcliffe Observatory in Pretoria, South Africa, Andrew Thackeray, had taken the limited data available on the star Eta Carinae and tried to estimate its distance and absolute magnitude. In the fifteen years since Enrique Gaviola had taken his pictures of Eta Carinae, Thackeray had been among a handful of scientists in the southern hemisphere who had periodically given the nebulous star a look, trying to understand its mysterious growing Homunculus.

Based on the size and expansion rate of Eta Carinae's outer shell of gas, Thackeray had estimated the star's distance from Earth to be from 3,000 to 9,000 light-years, with the larger number being more likely. At such a distance, Thackeray deduced that during its maximum eruption in 1843 Eta Car had reached an absolute magnitude of −14. At −14 magnitude Eta Car had been shining 34 million times brighter than the Sun, and had been able to hold that brightness for more than a decade. Even today, he estimated, the star was still shining at −6 magnitude, 21,000 times brighter than the Sun. Thackeray then concluded that the slow decline in Eta Car's brightness since the 1840s could be explained if Eta Carinae were either a "slow nova" or even an "ultra-slow supernova."[1]

Burbidge knew this conclusion couldn't be right. He had spent the last decade studying the birth, life, and death of stars. About the same time Thackeray had written his paper on Eta Carinae, Burbidge, his wife Margaret, William Fowler, and Fred Hoyle had written a landmark 100-plus-page-long paper showing that everything but the very lightest elements in the universe had to have been produced by the nuclear reactions inside stars. "[Thackeray's conclusion] was a contradiction in terms," Burbidge thought. "I was going around saying, well y'know, another typical observer who doesn't know a bloody thing about elementary theory."[2]

Thackeray's description of Eta Carinae, however, piqued Burbidge's interest. He began to take a close look at what was known about this strange star, and found its inexplicable behavior intriguing.

He reviewed the past research, going back to John Herschel. He looked at the images that Enrique Gaviola had taken in 1945. He studied the spectra that had been recorded of the star and its nebulosity. He talked about the subject incessantly with everyone, whether they wanted to listen or not.

As much as Burbidge doubted Thackeray's conclusions, Thackeray's research was enormously helpful. In 1951 Thackeray had spent nine nights getting an extensive set of spectrograms across the face of the Homunculus. Then, over the next two years he obtained the first infrared spectrum of the star. From these data, Thackeray had concluded that the shell of gas surrounding the star had been ejected from it, and was expanding outward at about 300 miles per second.[3]

To Burbidge, the data suggested something far more significant. The star's brightness and distance indicated Eta Car was a very massive star, sixty to a hundred times heavier than the Sun. The star's Homunculus and gas halo meant that Eta Car had been ejecting matter for a considerable period, and was still doing so at a prodigious rate. This also meant that the star was past middle age, and already beginning its dying process.

Taking these things together, Burbidge could not help but come to a startling conclusion. As he wrote in a 1962 paper on the subject, Eta

Carinae was "rapidly exhausting its supply of nuclear fuel, so that its evolutionary rate is speeding up exceedingly rapidly." He then added, "On the basis of this argument, we conclude that Eta Carinae is a likely candidate for becoming a supernova, though whether this will occur in the next 100 or the next 100,000 years we have no means of estimating."

Burbidge was hardly convinced that this was the right answer. Another theory proposed that Eta Car was a very massive protostar, more than 100 solar masses, so heavy that it was unstable under its own weight and was undergoing periodic contractions that caused it to eject material and brighten dramatically. Though the images and spectroscopy clearly indicated that the star was ejecting matter, neither were detailed enough to tell Burbidge or anyone else anything about how the star was doing it. Thus, choosing which theory was right was simply not possible.

Nonetheless, the behavior of this strange star was so unusual that any new information that could be gleaned about it would be invaluable.[4]

■ ■ ■

Even as Geoffrey Burbidge was puzzling over Eta Carinae, the entire world of space and astronomy was in an uproar for entirely different reasons. The launch of Sputnik had sent shock waves through the intellectual and political communities of the United States. In a rush to try to match the Soviets, NASA had been formed, the Mercury program had been instigated, and President Kennedy had declared that the United States was going to put a man on the Moon by the end of the decade.

The Burbidges weren't paying much attention to this uproar, however. They were observational and theoretical astronomers, focused on the sometimes difficult task of using ground-based telescopes to obtain data from the sky. "I wasn't really thinking of doing astronomy from space," Margaret remembered in 1984. "I was too busy trying to get some results from the ground."[5]

During the summer months, when not teaching, they had been struggling to use the 82-inch telescope at the McDonald Observatory in Texas, then managed by the same University of Chicago that operated Yerkes. In the early 1960s they had shifted their research from stellar evolution to studying faint and distant galaxies.

Then, in the spring of 1962, just after Geoffrey Burbidge had published his paper on Eta Carinae, the opportunity came to move to the newly opened campus of the University of California (UC) in La Jolla, north of San Diego. There they could work with the new 120-inch Lick Telescope.

UC, located in the suburbs of San Diego, was, unlike Yerkes, not an isolated place. NASA officials would repeatedly show up on campus to give talks on the wondrous possibilities of space astronomy. In one case, a NASA official had tried to excite his astronomer audience by noting that it would soon be possible to put the 120-inch Lick Telescope in space. Margaret Burbidge remembered him saying, "Just think how exciting that would be!" She also remembered the look of horror on the faces of her fellow UC ground-based astronomers, imagining their brand-new telescope torn from its moorings and flung wildly out of reach and into orbit.[6]

Of the NASA officials who came to San Diego to sell the idea of space astronomy, the most important for the Burbidges was a diminutive woman who had been NASA's chief astronomer almost from the day the space agency had been created.

Nancy Roman had always wanted to be an astronomer. Her father had been a geophysicist, and when she was ten he had gotten a job with the federal government, working from 1935 to 1936 in Reno, Nevada, before moving permanently to Baltimore, Maryland. It was during those two short years that Roman became hooked on astronomy, entranced by the clear night sky of the western desert.

She was so excited by the glittering heavens above her that during her second year in Reno the then eleven-year-old organized an astronomy club among her neighborhood girlfriends. "We learned the constellations, read astronomy, that sort of thing," Roman remembered in 1980. "After that I read almost everything [about astronomy] I could get my hands on."[7]

As the years passed and she went through the typical college and graduate school grind at Swarthmore and then the University of Chicago at Yerkes, Roman found herself increasingly trapped by school politics. Her main advisor at Yerkes, William Morgan, was a somewhat eccentric man who "had his hots and colds about people and subjects and everything else," Roman noted once. "There was a period of about six months in which he wouldn't even say hello to me when he saw me in the hall."[8]

Several astronomers tried to get her to switch to another university even before she completed her degree. She instead decided to stay on until she graduated, then find a position elsewhere.[9]

After she graduated, however, Morgan suddenly became more interested in her and suggested she remain at Yerkes as a research associate and help him with his own research. At the time he was focused on two research areas. First, he was doing a refinement of the Morgan-Keenan spectral classification system that he had worked out with astronomer Philip Keenan in 1950. This system, putting most stars into the spectral types of O, B, A, F, G, K, and M (and usually remembered by the phrase "Oh, Be a Fine Girl, Kiss Me"), is the basis of the stellar classification system that astronomers still use today. Second, he was working out the first accurate map of the spiral arms of the Milky Way galaxy.

The research associate position he offered Roman, however, was limited to a two-year term. Moreover, even when this ended and the school offered her a faculty position, it wasn't tenure-track. To keep her astronomical career alive she needed a tenure-track teaching position. Yet, she couldn't get any decent offers. "Morgan didn't want me to leave," she explained. "He saw to it that I didn't get any offers that would be very tempting."[10]

She kept her ears open, however, and in 1955 astronomer Gerald Kuiper let her know about a job opening doing radio astronomy at the Naval Research Laboratory in Washington, DC. Though she had never used radio telescopes before, she knew that radio astronomy was a new field that was going to revolutionize astronomy. More important, it would get her out of Yerkes and maybe get her career moving again.

Unfortunately, the office politics at NRL were no different than at Yerkes. When she arrived they introduced her to everyone, gave her a desk, and seemingly forgot about her. "After a few days, I decided they weren't going to give me anything to do." She dug out her old spectroscopic research from Yerkes and started working on that. Though eventually the others realized that she knew what she was doing and could be useful to the group, her work at the Naval Research Laboratory was never very fulfilling.

"It wasn't actually until after I left NRL that I found out what the trouble was," she explained in 1980. "They had had another woman, just about my age, also an astronomer, whom they had found absolutely useless and they were so happy when she left. Then to have another woman come in was a shock! . . . Here I was, thrown on them, and they didn't have the slightest idea what to do with me, except that they weren't particularly happy to have me."[11]

Had Sputnik not happened, Roman's astronomy career might have died here. However, Sputnik did happen, and with the formation of NASA in 1958 the man in charge of the space agency's new science program, Homer Newell, was looking for someone to head the space agency's astronomy program.

Throughout the late 1940s and most of the 1950s Newell had been a member of the Upper Atmosphere Rocket Research Panel, a semi-official group of scientists who met regularly and helped organize the distribution of available space for science experiments on the V2 as well as other suborbital rockets. During this time period he was a researcher at the Naval Research Laboratory in Washington, first heading the Rocket Sonde Research Branch, and later the science program for the Vanguard satellite project.

He was also, as described by John Naugle, a coworker and the man who replaced him at NASA when he retired in 1973, "a self-styled mathematician-turned-physicist. . . . He was an extremely hard-working, well-organized individual, but very touchy about his personal turf [and] frequently frustrated with himself and his subordinates because of his inability to cope instantly and perfectly with the requests and complaints of his superiors and scientific peers."[12]

When NASA was formed in October 1958, Newell negotiated the en masse transfer of almost fifty NRL scientists into NASA, forming the nucleus of the space agency's science staff at what was to become the Goddard Space Flight Center in Beltsville, Maryland. Newell in turn went to NASA headquarters in Washington, DC, where he was hired to run NASA's entire science program. He immediately began assembling a staff.[13]

Though she had been working at NRL, Roman had kept her distance from Newell's rocket group and had therefore not been part of the transfer to Goddard. "I sort of had the feeling that an awful lot of [the suborbital rocket work] was not hard science," she remembered in 1980. "I wasn't absolutely convinced that it was really doing good science, at that time."

Several months later, however, she was attending a lecture at NASA when John Clark, a former NRL scientist working with Newell at NASA headquarters, approached her. "By the way," he said. "Do you know anyone who would like to come and work for NASA and set up a program in space astronomy?"

For Roman, the chance to *head* the astronomy program at the new space agency was too good to be true. Unlike the suborbital work, NASA was proposing launching orbital research satellites with far greater capabilities. "The idea of coming in with an absolutely clean slate to set up a program that I thought was likely to influence astronomy for 50 years was just a challenge that I couldn't turn down."[14] Moreover, her radio work at NRL was still dissatisfying, since it seemed to require her to be more an engineer building radio receivers than an astronomer doing observations. In 1959 she accepted the job as NASA's chief astronomer, where it would be her task to help promote, design, organize, and build the world's first space telescopes.

Her task wasn't going to be easy. Prior to World War II, astronomers obtained the funding for their large telescopes from private individuals. When George Ellery Hale needed money to build the 40-inch refractive telescope in Wisconsin, he found Charles Yerkes, a Chicago financier, to pay for it. Then, when he wanted to build the 60-inch reflective telescope on Mount Wilson, he got Andrew Carnegie to fund it. Later he got entrepreneur John Hooker to pay for the 100-inch Hooker Telescope at Mount Wilson. Hale even found private funds from the Rockefeller Foundation for the 200-inch telescope on Palomar Mountain, even though it cost more than $6.5 million to build, or about $60 million in today's dollars.

Following World War II, however, the government became the primary backer of scientific research, especially for big projects like the launching of a space telescope. Getting government help, however, was far more complicated than convincing a lone eccentric millionaire to give you money. With government financing, no single person could simply say yes. Even though Homer Newell and Nancy Roman were eager to build a variety of space telescopes, they couldn't do it if they

Fig. 2.1. Nancy Roman, c1963, with a model of one of the Orbiting Solar Observatories. (Photo courtesy of NASA)

didn't get the wholehearted backing of many other people. First they needed the support of the public and Congress, who would finance the projects. Then, they needed the support of the astronomical community, who would propose, design, and use the space telescopes they would build. Finally, they had to garner support within the government bureaucracy itself on which projects to build. Any one of these interest groups could block construction.

In the late 1950s and early 1960s there was no problem getting public and congressional support. Sputnik and the space race had taken care of that. Similarly, bureaucratic support was easy, especially since Roman and Newell themselves were in charge of the astronomy program.

Getting the support of astronomers, however, was more problematic. Despite the passionate support of individuals like Lyman Spitzer, the overall astronomical community was still very skeptical about the practicality of space astronomy. Even though the 200-inch Hale Telescope had been available now for more than a decade, the most influential astronomers were not yet ready to move on to bigger and more ambitious projects. These individuals did not yet feel they had mined the Hale Telescope for all it was worth. Why spend vast sums of their limited resources on a project they did not yet need?

Most astronomers also saw the cost of building space telescopes as very excessive. For the cost of one space satellite, about $100 million, they thought they could build ten Hale Telescopes. Moreover, many feared that this vast investment for building a single space telescope might be wasted because of a single launch failure.

On top of these financial doubts, the technical challenges of building an instrument that could be aimed and used in space were daunting. Astronomers were very aware of the difficulties of building the 200-inch telescope. It had taken more than two decades to finish the Hale Telescope, including failures in casting both a smaller test mirror and the primary mirror. Then, once the primary mirror was successfully cast, it took an eleven-year effort to polish it into shape. The hurdles of building and using a space telescope would be far more challenging.[15]

Despite this overall skepticism, there were a few astronomers eager to launch instruments into orbit. Prior to Roman's arrival at NASA Lloyd Berkner, the head of the Space Sciences Board at the National Academy of Sciences, had sent out a letter to the astronomical community, soliciting ideas for space telescopes. The handful of respondents would all become the pioneers of space astronomy, including Spitzer, Arthur Code of the University of Wisconsin-Madison, Leo Goldberg, director of the Harvard College Observatory, Fred Whipple of the Smithsonian Astrophysical Observatory, and Albert Boggess of the Goddard Space Flight Center.

Unlike Spitzer, Art Code had not publicly lobbied for the construction of space telescopes in the 1950s, though he had been involved in the early U.S. V2 rocket missions. After Sputnik, however, he decided that this is what he wanted to do. "Astronomers dreamed of observing above the Earth's atmosphere for a century or so," he remembered in 1982. "It would be done whether there would be good astronomers involved or not. And so somebody better do it, and I decided I'd be one of those somebodies."[16]

For their first space projects Spitzer and Code both proposed launching an ultraviolet space telescope, with Code's doing broad survey work and Spitzer's aimed at more focused observations. Goldberg in turn suggested building an instrument for observing the Sun, while Whipple proposed a package of four small 12-inch telescopes, using television technology to produce ultraviolet images of various star fields.

To put these instruments into orbit the scientists, in conjunction with Newell, Roman, and others in NASA, came up with the idea of building a series of Orbiting Astronomical Observatories (OAOs). Each would use the same basic body and pointing and communications systems. They would also be somewhat large. As Art Code noted many years later, "[OAO]'s payload would be some 2,000 pounds of telescopes and could be a meter in diameter. And here we had this little 100-pound 10-inch telescope."[17]

NASA's solution to this excess of space was to cluster multiple instruments on each OAO, with the first OAO carrying Code's ultraviolet telescope as well as Whipple's package of high-resolution optical 12-inch telescopes. The second OAO would carry Al Boggess's experiment package plus several other instruments. Lyman Spitzer's proposal, a 32-inch ultraviolet telescope able to point at specific objects, required the most precision and was therefore placed third in the queue.[18]

As the OAO project developed, however, a number of astronomers became disenchanted with the OAO approach. They recognized that many experiments were incompatible and would be difficult to fly on the same satellite. (For example, after some discussion the solar observatory proposals were separated out and placed on a different series of satellites, dubbed the Orbiting Solar Observatories.) Moreover, OAO's large size required a larger rocket, which was expensive. Some astronomers thought it would be cheaper to launch multiple smaller instruments, using smaller rockets.

Unfortunately, the smaller rockets didn't exist, and even if they did, their price would not be cheaper. NASA's astronomy budget simply could not afford a lot of separate launches.

Other astronomers had more fundamental problems with NASA's astronomy program, complaining that Roman's approach was far too conservative. For example, astronomers like Aden Meinel at the Kitt Peak National Observatory were pushing for the immediate construction of much larger optical space telescopes, with a mirror at least 50 inches across, a project that was far too ambitious and risky for Roman to endorse in 1960. As she wrote several years later concerning another ambitious Meinel proposal, "Meinel's present attitude reminds me very much of his earlier attitude toward the 50-inch, 24-hour telescope. . . . I can not help but feel . . . that Meinel is letting his enthusiasm carry him overboard for more projects than he can successfully manage, particularly in view of his past reputation as a very poor administrator."[19]

Similarly, when Riccardo Giacconi of the private research company American Science and Engineering proposed in 1960 that NASA finance an x-ray instrument on a suborbital rocket, she turned it down, mostly because Giacconi's proposal said it would try to detect solar x-rays bouncing off the Moon, a phenomenon she knew was impossible. "If they had come to me to say they wanted to do a sky survey in x-ray, I think, admittedly in hindsight, that I would have supported them," Roman remembered in 1980. "I could not see supporting an experimental rocket to measure reflected solar x-rays from the Moon."[20]

When Giacconi's instrument was eventually financed and launched by the Air Force in 1962, however, he made history by detecting the first x-rays from outside our solar system. These x-rays, coming from a previously unknown object in the constellation Scorpio, originated from what we now know is a neutron star eating its binary companion.[21]

Nor were some astronomers complaining merely because they thought Roman was too conservative. Others, such as Jesse Greenstein, director of the California Institute of Technology's astronomy program, which then co-managed with the Carnegie Institute both the Mount Wilson 60-inch and 100-inch telescopes and the 200-inch Hale Telescope on Palomar Mountain, opposed her effort to create any kind of space astronomy program. He feared the money spent in space would siphon resources away from ground-based telescope construction.

Meanwhile, there were others who could not make up their minds. Though Leo Goldberg was a strong supporter of space astronomy, he could not decide whether NASA should launch a series of cheap small telescopes or the larger more ambitious projects, and over time would vacillate repeatedly between these positions.[22]

By 1962, there was great frustration among the astronomical community. While the manned space program was gearing up for an aggressive journey to the Moon, astronomers had not yet completed building their first satellite, with the first OAO launch still years away.

In an effort to garner better support from scientists as well as help them reach some sort of consensus about what they wanted NASA to do, Newell organized the first of what became a regular series of summer studies, bringing together all the important scientists in the field as well as members of NASA's bureaucracy to assess NASA's science program.[23] The first conference was held in the summer months of 1962 at the University of Iowa. More than a hundred scientists and almost a hundred NASA officials participated, forming joint working groups on such subjects as biology, lunar and planetary research, celestial mechanics, and meteorology.

The working group on astronomy, chaired by Leo Goldberg, included such space enthusiasts as Lyman Spitzer, Martin Schwarzschild, Fred Whipple, and Richard Tousey (who had obtained the first ultraviolet spectrum of the Sun using V2 rockets). Yet, even these astronomers couldn't make up their minds about what should be done. There was much argument about whether it was better to put a telescope in orbit or on the Moon. Others questioned whether the money was worth spending at all.

Spitzer proposed that NASA set up a committee to begin the planning studies for building a space telescope with a diameter of 100 inches or more. Armin Deutsch of the Mount Wilson Observatory, who was vice chairman of the working group, questioned this strategy. To him, it seemed foolish to begin planning such an ambitious project before NASA had even launched its first successful Orbiting Astronomical Observatory.

Nor was Deutsch alone in his doubts. Tousey was also unconvinced. Even Nancy Roman considered Deutsch's position reasonable. "I think it was just a case of my dragging my feet," she remembered in 1980. "I was positive that technologically we just were not ready."[24]

Roman's concerns were not trivial. Though by 1962 both NASA and the military had successfully flown a variety of observational spacecraft, from weather to surveillance satellites, the technical capabilities needed to make a space telescope work were far more difficult and challenging to achieve.

For example, in 1961 the weather satellite *Tiros 3* took the first space pictures of the Earth, beaming back pictures of Hurricane Betsy in September 1961. The Earth is very bright, and only a split-second exposure was required for *Tiros 3* to get an image. The satellite did not have to stay locked on its target in order for the image to be clear and sharp. Galaxies, quasars, and planetary nebulae, however, are very small and faint. For a space telescope to photograph them it would have to be able to stay on target continuously for hundreds, even thousands of seconds, even as it was orbiting the Earth at 17,500 miles per hour. In 1962 no one had any idea how to build such a precise pointing system.

Moreover, even if a space telescope could be pointed, in 1962 no one knew how to get the data from space. The military's surveillance satellites used film, which was returned to Earth and recovered by snatching the descent capsule out of the air with an airplane. This technique was expensive, and required more capabilities than were available to Roman's meager astronomy budget.

The first weather satellites had used television signals, reducing the data into a series of coarse line scans, which could be beamed back to Earth. For space astronomy, however, this technology would not work. Pictures of hurricanes and storm systems did not require high resolution. Faint galaxies and planetary nebulae required extremely sharp images, far sharper than anything yet achieved even on the ground.

Faced with these serious technical issues, the working group was unwilling to accept Spitzer's proposal. While its final recommendation said that NASA should plan to build larger telescopes, it also included Deutsch's minority view that it was simply too early to proceed. "At a time when not a single image of a celestial body has been obtained in a satellite," the final report said, "it is premature to . . . study a space telescope larger than the [not yet launched] 38-inch telescope of the [first] OAO."[25]

This inconclusive recommendation made it difficult for NASA to move forward. Lyman Spitzer's dream of building the first optical space telescope was once again put aside, pending further developments.

Over the next three years, however, there were enormous advances in space engineering, epitomized most by the success of the manned Gemini program. First proposed in 1961, the program had quickly ramped up, and by March 1965 it was putting two men in orbit every two months. By the summer of 1965 the program had already successfully tested precise maneuvering in space as well as completed its first spacewalk. In addition, NASA's science program had successfully launched three Ranger probes to the Moon, taking thousands of photos that were successfully transmitted back to Earth before the spacecraft crashed on the lunar surface. On top of this, the Mariner 2 spacecraft had flown past Venus in 1962, sending back data on the cloud-covered planet's hot atmosphere, while the Mariner 4 spacecraft had been launched toward Mars in 1965, and had been transmitting back detailed information on the solar wind as it traveled to Mars during the spring.[26]

When NASA scheduled another study conference for the summer of 1965 at Woods Hole, Massachusetts, the social atmosphere was far different than at the Iowa conference. The circumstances more pleasant, located as it was on Cape Cod by the ocean, where scientists could rent cottages, bring their families, and even stroll down to the ocean during lunch breaks to go swimming. Besides, the successes in space during the past three years had helped ease the doubts of many astronomers. "At the time the future seemed unlimited," remembered Leo Goldberg in 1978. "In 1965 the sky was the limit."[27]

This time it was Lyman Spitzer who chaired the working group on astronomy, with fellow space enthusiasts Art Code, Fred Whipple, and Aden Meinel forming the heart of the committee. This time the focus was on getting a large space telescope built, and to do it in conjunction with the manned space program. As the committee recommended in its final report, "A single large telescope, maintained at regular intervals [by humans], could have a useful scientific life of decades, as compared with about a year or less for complex unmanned equipment at present."

The report then went on to unequivocally call for the construction of a 120-inch diameter space telescope, "requiring the capability of man in space." They also noted that such a telescope maintained by humans was "becoming technically feasible and will be uniquely important to the solution of the central astronomical problems of our era."[28]

The astronomers were in such agreement at Woods Hole about building a big space telescope that at one point things devolved into a lighthearted discussion on what to name it. Fred Whipple suggested that they name it the "Great Orbiting Device," or GOD.

Leo Goldberg objected, saying such a name wouldn't go, because senators think of themselves as God. "You don't want to tread on senators' toes."

Then someone suggested the name GOT, for Great Orbiting Telescope. Once again Goldberg objected, "That's just an obvious contraction of Gott [God] in German and the senators won't buy that."

They finally ended up calling it LST for Large Space Telescope, though even with this name there were some minor complaints. "They thought somebody might misread this as the Lyman Spitzer Telescope," astronomer Laurence Frederick remembered. "I thought to myself, so what? He's been the prime mover."[29]

Despite this air of assured confidence, there were still some doubts. "The recommendations . . . were very ambitious," noted Goldberg. "It wasn't clear how you went about designing and operating a large space telescope."[30] But whereas at the Iowa conference such doubts prevented the formation of a committee to begin design research work, this time the doubts served to get a committee started. The working group not only agreed to form what they called the Large Space Telescope Ad Hoc Committee to look into building Spitzer's giant space telescope, it made Spitzer chairman of that committee, and included Art Code and Nancy Roman as members. Though Roman still had worries, she now agreed the time was right to move forward. "We were never going to get there if we didn't start somewhere, and this was as good a place as any to start," she noted in 1984.[31]

The consequences of the Woods Hole summer study were far-reaching. The report, over six hundred pages long, was not just about space astronomy, and it made recommendations in every area of space research, from biology to lunar and planetary exploration. To implement these recommendations, NASA administrator James Webb put together a committee in 1966 under the leadership of Norman Ramsey, professor of physics at Harvard University. On that committee were two very prominent astronomers: Martin Schwarzschild, at that time building Stratoscope II, a 36-inch upgrade of his first balloon telescope, and Leo Goldberg.

Like Spitzer, Goldberg was an important figure in the politics of government and university science in the post–World War II period. His background was, however, completely different. The son of poor Polish immigrants, he had spent his early childhood years growing up in the Jewish ghettos of Brooklyn. His father had been a hatmaker, and his mother had died in an apartment fire when he was nine. When he was twelve they moved to New Bedford, Massachusetts, where his father opened a men's hat shop.

As poor as they were, he had been raised with a deep love of learning, and had therefore gotten the highest grades of any boy in his high school class, including being crowned spelling champion of southeastern Massachusetts. As a result, he earned a scholarship to the Harvard Engineering School. The scholarship, however, only paid for his tuition. "I had no money whatsoever."[32] To pay his living expenses, one of the members of the scholarship committee offered to loan Goldberg whatever additional money he needed at no interest. Goldberg also arranged to spend his summers working for a hotel on Nantucket Island.

"I always had the notion that I wanted to do something that really turned me on," he remembered in 1978. As a city boy, he had never seen the night sky in all its glory, and had had no exposure to or interest in astronomy. On Nantucket, however, the night sky glistened. Moreover, the hotel owner was an amateur astronomer, and would take Goldberg to lectures and open house nights at the local Maria Mitchell Observatory, which had a 7.5-inch telescope. Near the end of the summer after his sophomore year, a total solar eclipse rolled across Nantucket Island. The hotel owner and Goldberg went up onto the sloping hotel roof where the owner took pictures with his Brownie camera and Goldberg took notes.

Still, Goldberg had no plans to change his major from engineering. When he got back to school, however, he arranged to drop the calculus course he was taking so that he could take it with a better teacher one year later. In looking for a substitute, he picked *Astronomy I* out of curiosity. The course was being taught by astronomer Bart Bok, who was known to be a very inspiring lecturer. "By the end of that year, I decided I wanted to be an astronomer."[33]

From this moment on, his path into the elite world of astronomy was like greased lightning. By the time he was 33 he was head of the astronomy department at the University of Michigan. In subsequent years he became involved in a whole range of government committees, from helping to found the National Science Foundation to becoming president of the International Astronomical Union. After Sputnik, he immediately got involved in NASA's first efforts to do space astronomy, joining the Space Science Board of the National Academy of Sciences and building several instruments for NASA's series of Orbiting Solar Observatories.

By 1966, however, Goldberg was very unhappy with how NASA had been running its space astronomy program. Worse—and significant for the future of Nancy Roman—he believed that the program was too timid under her administration. "We started out at Harvard dealing with Nancy Roman," he remembered in 1978. "That was disastrous," he added in his brusque, streetwise manner. "She had no imagination whatsoever."[34]

In fact, Roman's central position as the arbiter of all American space astronomy had increasingly made her a focus of dispute both inside and outside NASA. Throughout the early sixties, while she spent endless hours selling the space program to uninterested astronomers, she organized and approved a variety of small space research projects as well as some larger ones. As Margaret Burbidge remembered, Roman "was doing her best to get all the astronomers that she knew interested in working from space."[35] Nonetheless, most of her work was tedious paperwork and administrative tasks, not research, which hurt her reputation with many astronomers. Worse, her hard-nosed and realistic manner of approving or denying research projects had made her disliked by many in the astronomical community. "She would tell people home truths to their face in undiplomatic language, and she didn't do herself well by doing that," noted astronomer Richard Henry. "She should have been a little more greasy, but she wasn't, she was really quite straightforward."[36]

Under Goldberg and Schwarzschild's influence, the Ramsey committee recommended that NASA's astronomy program be shifted out of NASA and be placed under the control of an independent university consortium. Throughout 1967 these two men met repeatedly with Roman and her superiors at NASA, arguing that NASA's space astronomy program should be run by astronomers, not NASA.

Though Newell and Roman were willing to listen and accept many of Goldberg's and Schwarzschild's recommendations, they were unwilling to accept the idea of shifting control of the program out of NASA. It was their responsibility to determine how the taxpayers' money was spent. By law, they could not give that control to outsiders.[37]

These parleys did, however, result in the formation of the Astronomy Missions Board, chaired by Leo Goldberg. Formed in the fall of 1967 and meeting almost two dozen times over the next three years, the board was designed to replace the periodic summer studies with a continuous and independent committee providing advice from non-NASA astronomers on what NASA's astronomy program should do next.[38]

In the end, the recommendations of the Missions Board had a limited effect on the kind of astronomy projects NASA would approve in later years. In terms of the Hubble Space Telescope, however, the board ended up being pivotal, in that it introduced one man to the idea of an optical space telescope who would end up becoming one of the most important figures in Hubble's history.

■ ■ ■

As far as C. Robert "Bob" O'Dell can remember, he had always been fascinated with the heavens. Growing up in the open flat plains near East St. Louis, Illinois, the glimmering and crisp night sky had always beckoned him. When, as a twelve-year-old in 1949 he had been asked to write a one-page composition on what he wanted to do when he grew up, he wrote how he wanted to be an astronomer at the observatory on Palomar Mountain, using the just completed 200-inch Hale Telescope.

Nor were the stars the only things drawing him upward. During these same childhood years he lived right next door to the airfield of what was then the Parks College of Aviation. As he told me, "Our backyard was the fence of the airport." Planes would buzz the house. Later, after the war, his family boarded the spouses of aviation students who were training at Parks under the GI Bill. "I was always around airplanes," he noted. In high school he managed to arrange his first introductory flight and finagled that into his first lesson. "It became my first love." By the time he was seventeen he had flown his first solo flight, getting his pilot's license one year later.[39]

O'Dell was a restless and hardworking man who couldn't help filling every second of his life with some activity. "I was doing something all the time."[40] In his freshman year in college he not only attended classes, worked on the side to pay his bills, and joined the college's amateur astronomy club to use the school's 12-inch telescope, he also bought a percentage in a plane so that he could keep flying.

Fig. 2.2. C. Robert O'Dell. (Photo courtesy of Ed Weiler)

Initially, he chose aviation as his life's goal, majoring in aeronautical engineering at the University of Illinois. Unfortunately, he found the work dry and boring. Later he remembered with humor how "aeronautical engineers weren't very lovable people." In describing his dissatisfaction with engineering to one of the university's astronomers, Stan Wyatt, O'Dell proposed switching from aviation to astronomy. Wyatt told him that he could become an astronomer if he wanted, but if he did, he should expect to be poor his whole life. "You'll always have to drive a beat-up car and you'll never own your own home," Wyatt explained.[41]

Still, astronomy seemed far more interesting than engineering, and O'Dell decided to make the switch.

O'Dell's change of major from engineering to astronomy, like Goldberg's, quickly shifted his career into high gear. He got his college degree in three years while working numerous odd jobs to pay for school, including being a cook in a women's dormitory and helping to embalm corpses at a local funeral home. Then he took only three years to get his PhD at the University of Wisconsin, the heart of space astronomy. Here, Art Code was just starting to put together his first ultraviolet space telescope for Nancy Roman's OAO program. Then O'Dell got offers to teach at both Berkeley and Michigan, and grabbed the job at Berkeley, where he divided his time between researching planetary nebulae and running a project to find a southern hemisphere location for a new telescope to be built by Berkeley. In fact, less than sixteen years after he had written his sixth grade composition dreaming about being an astronomer using the 200-inch Hale Telescope, he had actually made it happen. Things couldn't have been better.

Then they got better. O'Dell was offered the job as chairman of the University of Chicago astronomy department, owner of the world's largest refractive telescope, the 40-inch Yerkes telescope at Lake Geneva, Wisconsin. He took it.

By 1966 O'Dell was at the top of the astronomical world. Only twenty-nine years old, he was head of one of the most prestigious astronomy departments in the world, doing research with some of the world's biggest telescopes, and having the time of his life.[42]

Others quickly noticed him. In 1967, when Leo Goldberg was arguing his case before NASA, he invited Bob O'Dell to join him at a meeting with the director of the Goddard Space Flight Center. Throughout the early 1960s Goddard had developed a poor reputation among astronomers, whether deserved or not, for the quality of its scientists. The goal of the meeting was to help the Goddard Center figure out ways to attract better scientists to work for it.*

O'Dell, still only thirty years old and by far the youngest man in the group, was honored by the invitation and the respect he received from these bigwigs of astronomy and space exploration. It stoked his ego, making him feel a member of the small elite group that made some of the nation's most important policy decisions.

When Goldberg formed the Astronomy Missions Board two months later, he once again asked O'Dell to be a member. "He kind of adopted me," O'Dell remembered in 1982. "I think he saw a young Leo Goldberg. . . . I was the youngest person on the . . . Board."[43] The two men were similar in temperament. Both could be bluntly honest if they felt it necessary. Both were sharp and outspoken. Both wanted power and the ability to influence national space policy for decades to come. "I

* One of O'Dell's recommendations, Jack Brandt, was eventually hired, becoming in the 1970s and 1980s one of Goddard's most respected space scientists. In fact, Brandt became the principal investigator for the High Resolution Spectrograph, one of Hubble's first instruments.

was very much into power," O'Dell remembered in 1985. "I loved being first."[44]

At the time, however, O'Dell knew very little about space astronomy. "I was too much just caught up with the grand and glorious idea of doing [ground-based] astronomy."[45]

Joining Goldberg's Astronomy Missions Board soon changed that. From late 1967 to mid-1970 the board met twenty-three times, almost monthly, forming subcommittees on optical, infrared, radio, and ultraviolet astronomy in space, as well as a subcommittee on ground-based astronomy (the recommendations of which O'Dell eventually wrote). For the optical panel the Missions Board incorporated Lyman Spitzer's Ad Hoc Committee on the Large Space Telescope in toto, with Spitzer presenting his committee's findings in person. The report, entitled "The Scientific Uses of the Large Space Telescope," was later dubbed by everyone who read it as "Chairman Spitzer's Little Black Book," a nickname inspired by Chairman Mao's own "Little Red Book," which was in the news at that time.

Though O'Dell had been unaware of space astronomy beforehand, he very quickly got caught up by the excitement of this new uncharted territory.[46]

Of the many Missions Board meetings he attended during those three years, the fifth was by far the most important in changing his life. Held for two days in March 1968 in Boulder, Colorado, it was here that O'Dell met Wernher von Braun for the first time. "The great man, the great Hun," as O'Dell described him years later.[47]

Von Braun, then director of the Marshall Space Flight Center in Huntsville, Alabama, had been invited to give a presentation describing Marshall's future plans for space exploration. Von Braun also came looking for work. After almost a decade of effort, the first lunar landing was only a little more than a year away. Unless von Braun could garner new projects for Marshall, the center faced layoffs or even shutdown.

O'Dell couldn't help being impressed with von Braun as he tried to sell Marshall to the astronomers. "You know, no one could be more dynamic than Wernher as a speaker," O'Dell later said. "They had this beautiful pitch, with Wernher speaking and Ernst Stuhlinger flipping charts for him and keeping his boss smart so he didn't put his foot in his mouth." Stuhlinger, head of Marshall's space science laboratory, had come along to help von Braun convince the scientists that Marshall could do science as well as manned space exploration. Together they wowed the scientists. After they had left, O'Dell remembered one astronomer saying, "Wow! Gee whiz! Wasn't that great?" while another noted, "Yeah, he used to get his practice giving presentations to Hitler."[48]

Von Braun in turn was duly impressed with O'Dell. At lunch, the two men sat together, and as they ate von Braun described how he was looking for scientists to do consulting work for Marshall. "I want some people to give me advice," von Braun explained to O'Dell. "Would you consider something like that?"

Though at first ambivalent, O'Dell couldn't resist the opportunity to help shape Marshall's research program. Over the next few years he, along with a handful of other scientists, made several trips each year down to what O'Dell liked to call "Hunts Patch, Alabama," giving von Braun advice on the kinds of space research astronomers wanted. "To von Braun's credit, he took that advice," O'Dell said. "He went, you know, where the business lay, because he was trying to insure the future of his organization. . . . By the time he left [Marshall], it was one-third the size it was at the peak, and he was just desperate to get worthwhile things going."[49]

By 1971, in response to the advocacy of the Astronomy Missions Board as well as Spitzer's committee, NASA was increasingly ready to commit itself to the construction of a large space telescope. Under Nancy Roman's leadership, a Large Space Telescope Science Steering Group was formed to begin the hard work of actually designing the telescope. The agency had already commissioned studies of the concept from a host of private contractors, including Boeing, Grumman, Perkin-Elmer, Itek, and Martin-Marietta. It had also run a major optical telescope workshop at Marshall in 1969, where more than eighty technical engineering papers were presented. Since the space agency was then also doing its first design work on the space shuttle (which was eventually approved in 1972), most of these early space telescope studies incorporated the shuttle into their design, with the shuttle launching the telescope and returning repeatedly to do some form of maintenance and servicing.[50]

Both the Goddard Space Flight Center and the Marshall Space Flight Center had completed their own telescope design studies as well, with both increasingly in agreement on how the large space telescope should be built. With the Apollo lunar program winding down, von Braun very specifically wanted Marshall to get the job of building the telescope. Giving von Braun's center this job, however, was not a straightforward decision for the NASA officials who ran the agency. Since NASA's inception Marshall had done very few science projects. Instead, Marshall had been the place that built the Saturn 5 rocket and the engines that sent it to and from the Moon.

Almost all of NASA's astronomy work had been designed and built at the Goddard Space Flight Center in Beltsville, Maryland, just outside Washington, DC. Though building a space telescope seemed a natural for Goddard, and though many of its scientists and engineers eagerly wanted the job, John Clark, Goddard's director throughout the 1960s and early 1970s, was not interested. Instead, he wanted to diversify the center's work, getting it more involved in what NASA bureaucrats called "applications," engineering work that the agency used immediately to explore space. He along with many headquarters officials also thought Goddard didn't have sufficient staffing to do the project, and could see no way of shifting staff from elsewhere to make up the shortage.

Jesse Mitchell, who was then director of the physics and astronomy program at NASA headquarters under Newell, weighed the options and realized he was faced with a Hobson's choice. The early 1970s was a depressing time for NASA. Having won the space race to get to the Moon, NASA was faced with shrinking budgets and a public and government increasingly unwilling to finance further space adventures. Mitchell simply didn't have enough new projects to give to both Goddard and Marshall. If he chose Goddard to build the space telescope, he risked the possibility that Marshall would die for lack of work. Politically, that would not go over well with certain members of Congress in Alabama. In turn, if he chose Marshall, he was faced with the same possibility at Goddard, since the so-called "applications" work that Goddard wanted simply did not exist. Rather than choose, Mitchell wanted to follow that sage advice of former Yankee baseball catcher Yogi Berra, "When you come to a fork in the road, take it." For Mitchell, the best decision would be to split the job of building the space telescope, giving both centers work in order to keep them both alive.[51]

However, just as astronomers like Goldberg and O'Dell did not trust the quality of the work out of Goddard, they also had little confidence in Marshall. Marshall had run the science program flown on Skylab, including the Apollo Telescope Mount, which had been designed as a space telescope for studying the Sun. Leo Goldberg was heavily involved in that project and had been unhappy about the bureaucratic battles he had experienced.

For example, in the late 1960s, the Harvard College Observatory, where Goldberg was then director, was building a very sophisticated ultraviolet spectroheliometer for studying the Sun, to be attached to the Apollo Telescope Mount. In 1967, however, George Mueller, head of the manned program, came to Goldberg and told him that NASA wanted to accelerate Skylab and have it up by 1969. Could Goldberg get Harvard to quickly build a smaller, less complicated instrument instead? Goldberg thought about it, and realized that since the solar maximum would occur in 1969, it would be worthwhile to have a simple flare spectrograph in space at the time to study the maximum's many flares. He therefore agreed, on the condition that Harvard could still finish the more sophisticated spectroheliometer for later flight.

Not surprisingly, Skylab kept getting delayed, until Goldberg realized that it was going to launch during solar *minimum*. The flare spectrograph would be pointless. He called Mueller and explained that if they stayed with the spectrograph "we'll be the laughing stock of the astronomical community. We want to go back to our original experiment."

He was told, no. NASA no longer had the money to finish the other instrument. Fly the flare spectrograph, or nothing.

Goldberg chose nothing, telling them in writing that he was formally pulling Harvard from the project. "They came right back and said 'Now, let's talk this over.'" Goldberg got the original experiment.[52]

Since no astronomers worked at Marshall, the astronomers worried that this kind of scenario would be repeated with the space telescope, that as the project matured their interests would simply not be championed. Worse, because Marshall was located in Huntsville, Alabama, a relatively isolated and rural area of the south, it seemed unlikely that any respected astronomer would be willing to come there to look after their interests.

In the summer of 1971, at a meeting of Nancy Roman's Large Space Telescope Science Steering Group held at Huntsville, Alabama, this issue was argued at great length. Again and again the astronomers repeated their doubts: Without an astronomer of some prestige in charge of the project at Marshall, there was little chance the astronomical community would support having Marshall build the telescope.

That evening, Stuhlinger invited the group to his home for a barbecue. The house, located on the side of Montesano Mountain on the east side of Huntsville, had a beautiful view overlooking the city from its backyard. As people watched the Sun set behind the city, Stuhlinger went up to Lyman Spitzer to talk about Marshall's problem. "What we need for this project is a top-notch astronomer who is willing to come to Huntsville for some time . . . and to be the center of space telescope activities as a scientist. It must be a man who has the confidence of the astronomers, who is an astronomer in his own right, who believes in the LST."

Spitzer agreed. Even if NASA chose to let Marshall build it, without such a person the project would end up being opposed by too many in the astronomical community, and very likely fail.

Stuhlinger then asked, "Wouldn't you like to do that? Could we invite you to become a member of our force here, a guest, and be with us?"

This was the offer Lyman Spitzer had been working for since 1946. As early as 1947 he had been talking about his desire to do space research. In that year, when he took over as director of the Department of Astrophysical Sciences at Princeton, he had asked his colleague and friend Martin Schwarzschild to leave Columbia University and join him in running the department. Schwarzschild wanted to work with Spitzer. He liked him personally, and also trusted him to run the department to his advantage. Also, he was unhappy at Columbia. He and his wife were avid bird-watchers, and did not like the urban environment. "To live in New York, I really hated," said Schwarzschild in 1977. "To get outdoors from Manhattan was so difficult."[53]

Schwarzschild, however, was worried whether he could depend on Spitzer to stick with Princeton. So he wrote Spitzer a letter, asking him how long he intended to commit himself to Princeton.

Spitzer's response in 1947 was very concrete. He explained to Schwarzschild that there was only one circumstance that would tempt him to leave Princeton. "If [the United States] goes into a space program, and if to participate in astronomical research in the space program the setup required that one had to be at a government lab, then Lyman stated that 'I will join the national space program and leave Princeton.'"

Schwarzschild thought about this, and then decided he had nothing to worry about. "Lyman is foresighted, but he is also a dreamer," he thought. "I did not believe that [such a program] would happen in my lifetime."[54]

Now, almost a quarter of a century later, Spitzer was being offered that exact opportunity.

Things had changed, however. He was in his late fifties, and nicely settled with his family in Princeton, where he was only a few hours away from the Catskill Mountains and his weekend rock-climbing activities. He had good research facilities, the respect of colleagues, and a generally comfortable life. Moreover, he did not have to come to Huntsville to participate in the space telescope project. For the last several years Nancy Roman had already been providing him funds for Princeton to develop the camera and detectors that would be used on the telescope. He could stay in Princeton, build the camera, and still be part of the project.

Spitzer looked at Stuhlinger in his mild way and shook his head. "No, I don't think so. However, I know somebody who can do it."

Stuhlinger, though disappointed by Spitzer's refusal, immediately leaped at this. "Who is he?" he asked eagerly.

Spitzer pointed across the yard to a thin balding but young man in a turtleneck, standing next to a large tree. "That man over there. His name is O'Dell."

Stuhlinger's first thought was that there was no way O'Dell would consider the job. He told Spitzer, "I doubt he will come to the civil service [here in Huntsville]."

"Well, I am not so sure whether you are right." Spitzer shrugged. "Let's talk to him."

Spitzer and Stuhlinger walked over to O'Dell. "Hey Bob," Spitzer said. "Wouldn't you like to come to [Marshall] and be the LST man for scientists for a year or so?"

To Stuhlinger's surprise O'Dell thought for a second, and then said, "That sounds interesting."

O'Dell thought the idea was more than interesting. By this time, "I was caught up with the romance of space telescope," he remembered in 1982. "After all, that was going to be the most important instrument in astronomy during my lifetime, and the chance to be a key part of that was very attractive."

O'Dell was also restless in his position as head of the astronomy department at the University of Chicago. He was only thirty-four, and was looking for a change. His wife wasn't happy living at the Yerkes Observatory. Like the Burbidges before her, she found the place too isolating.[55]

Once Jesse Mitchell knew that a scientist of O'Dell's prestige was willing to consider moving to Marshall to head the space telescope project, he felt free to recommend Marshall for the lead position in building the telescope, with Goddard as support. On May 5, 1972, the decision was announced. The Marshall Space Flight Center was to be the "lead" center, designing the telescope structure and supervising its construction. Goddard in turn was put in charge of building the scientific instruments that would go inside the telescope body.[56]

By the end of that summer Bob O'Dell had taken a three-year leave of absence from the University of Chicago and moved to Marshall to become the project scientist for the Large Space Telescope. If all went well, he hoped to get the project funded, under construction, and launched by 1977.

What he didn't know was that it wasn't going to take three years, or even six, to do this. In fact, O'Dell's time in Huntsville was to last more than a decade. And even then it would be another nine years before the telescope was launched.

Meanwhile, for Lyman Spitzer, the decision to pass on the space telescope job was the first of several crucial events that would seal his own fate relating to the Hubble Space Telescope. By choosing the easier path of staying in Princeton, he had unwittingly begun the process that would eventually make him an outsider looking in.

3

Getting Money

The climber was tall, thin, and wiry. He moved smoothly, carefully, and methodically, working his way up the wall without pause. Two hundred and fifty feet below him stretched the green and shimmering meadows of the Yosemite Valley.

On a ledge beneath him, forty-three-year-old astronomer Ralph Bohlin fed out the safety line, watching intently to make sure he was ready to catch the man should he fall. Bohlin was especially concerned because the climber was not merely another human being, he was sixty-six-year-old Lyman Spitzer, director of Princeton University's Department of Astrophysical Sciences, one of the country's most eminent scientists and considered by many the father of the Hubble Space Telescope.

"I felt a little guilty [since] I usually let him lead on the harder parts," remembered Bohlin.

Spitzer's mountain climbing hobby had become exceedingly more sophisticated since the 1950s. In the 1960s Don Morton introduced him to more modern technical climbing techniques, and they as well as fellow Princeton professor Ted Cox would go on regular weekend trips to climb the cliffs in upstate New York. Morton and Spitzer also took part in several major mountain-climbing expeditions to some of the more remote places on Earth.

For example, in 1965 they were members of the second Canadian expedition to Baffin Island, where they climbed two mountain peaks, including the first ascent of 5,500-foot-high Mount Thor. At one point Spitzer hesitated, daunted by the "vertical rock face above us." Morton wanted to go on, however. As Spitzer wrote later, "He had sort of counted on getting me to the top, and after eating a little supper of cheese, apricots and chocolate I felt more daring and decided to go on up. I am very glad I did, since the climbing was really not so terribly difficult. We got to the top at midnight and had a fine view of the entire range, with the sun only just a bit below the horizon."

Since Spitzer had to return home sooner than the others, his plan had been to leave Morton after a second climb and hike solo twenty miles to the coast to meet a local Eskimo with a boat.

Unfortunately, the two missed each other. If Spitzer didn't get 20 miles further down the coast to the village of Pangnirtung by the next day, he would be left stranded on Baffin Island. With no spare food or supplies and no way to contact anyone if something should go wrong, Spitzer started out, fording several roaring and freezing streams along the way. "In retrospect we had taken an awful risk," Morton wrote years later. "Nothing serious happened thanks to Lyman's stamina and skills in wilderness travel."[1]

Another time—in a more offbeat adventure—the two astronomers decided to climb the Gothic tower of the Princeton Graduate College. This ornate structure had been climbed by students and teachers in the past, but Spitzer had never done it and wanted to. One of Spitzer's graduate students provided the key for the building, and early one Sunday morning Spitzer, Morton, John Wrigley, and Michael Shull proceeded to set up a top rope so that they could be safetied as they climbed.

Unfortunately, no one had told the campus security people, who showed up as Spitzer was about two-thirds of the way up the side of the tower. "[They] demanded that we cease and desist, threatening to call the town police," wrote Morton years later.[2]

Now, in 1981, Spitzer was in Yosemite with Bohlin, and—fearless as usual—he had climbed too far above his protection. Unable to rest, he got tired, lost his grip, and fell.

Below, Bohlin reacted immediately. "I started to pulling in the slack." Then the chockstone protection that was lowest and nearest to Bohlin came loose, which frightened him and made him hold off, letting Spitzer's fall take the slack to avoid putting more stress on the remaining chocks.

This was a mistake. The slack in the line allowed Spitzer to drop enough to hit the ledge where Bohlin was sitting, about twenty feet to the side. "To me it looked like [I] hadn't really broken his fall that much.

I was greatly relieved to see him move and greatly relieved to find that it was only a broken arm."

Even with a broken right arm Spitzer was unperturbed. Carefully they worked their way down the mountain, with Bohlin first lowering Spitzer and then following. When they reached the ground Bohlin suggested they head to the hospital. "No," said Spitzer in his genteel way. "We're going to have dinner first." They put his arm in a sling, made a nice dinner at camp, and then went to the hospital.

This accident was in many ways the swan song of sixty-six-year-old Spitzer's rock-climbing career. Though he was climbing again in about a year, "he didn't climb a whole lot after that," noted Bohlin.[3]

Spitzer's fearless and adventurous spirit was not unusual among American astronomers in the postwar era. For example, in 1967 when Bob O'Dell was director at Yerkes, he had joined Spitzer on a climbing expedition to central British Columbia. Unlike Spitzer, Bob O'Dell didn't find mountain climbing to his taste. "I was never a good rock-climber." (His strongest memory of this trip was how the expedition ran out of food early.) Then, in 1976, when the space telescope project was in full swing, the two men had arranged to go diving together in the giant tank at the Marshall Space Flight Center where astronauts were simulating a spacewalk using mock-ups of the shuttle and the space telescope.

Flying remained O'Dell's first love. Since moving to Huntsville he had gotten passionately involved in competitive aerobatic flying. He purchased a small plane in 1973, and modified it so much himself that the FAA considered him the builder. By 1980 he knew so much about the sport that he even penned the book *Aerobatics Today*, considered for many years the sport's textbook.

Meanwhile, the space telescope project was slowly inching forward. When O'Dell arrived at Marshall on September 1, 1972, to take over as project scientist, he immediately found himself embroiled in a turf war between Marshall and Goddard, a situation that was to be repeated innumerable times in the years to follow.

The working relationship between the two centers had supposedly been worked out earlier that summer, with Goddard handling the science instruments under Marshall's leadership. O'Dell discovered that this agreement was so vaguely worded that it was causing confusion and discord. Moreover, the deep resentment felt by Goddard managers because

Fig. 3.1. Bob O'Dell and his airplane, c1979. (Photo courtesy of C. Robert O'Dell)

the center had lost the space telescope project to Marshall caused them to give it a low priority. "I think it's fair to say that they did not put their very best people on the program," noted Robert Bless, principal investigator for the telescope's High Speed Photometer being built by the University of Wisconsin.[4]

Goddard managers also resisted giving up control of the project. For example, the Goddard manager in charge of science instrument development, Ken Hallam, wanted all scientific work, including all contacts with the scientific community, to go through him, not O'Dell. On top of that, Hallam was still insisting on using Goddard's design for the telescope, something that neither Marshall nor O'Dell wanted.

In a series of brutal meetings throughout the fall of 1972, dubbed by O'Dell "war councils" in his diary, O'Dell had Hallam removed as instrument scientist and negotiated a new working arrangement with Goddard. O'Dell got full control, and would work with George Levin, whose job would be limited to managing construction of the science instruments as designed under O'Dell's leadership.[5]

The telescope project was hardly in O'Dell's control, however. Unbeknownst to him, shortly thereafter an event much more significant to the future of the space telescope took place at NASA headquarters. The occasion was a December 21, 1972, meeting with NASA administrator James Fletcher. One of the disagreements between the Marshall and Goddard managers at the time had to do with the use of contractors to build the telescope. Marshall managers wanted to hire two associate contractors and manage them as equals. One would build the telescope itself, essentially a frame with two mirrors precisely positioned. The other would build the telescope's support systems module, containing the power supply, pointing system, communications equipment, and other equipment for keeping the telescope operating. Goddard managers instead advocated hiring a single prime contractor to build everything, who would in turn hire subcontractors to build whatever sections the prime contractor couldn't handle.

The December 1972 meeting with Fletcher, to which O'Dell was not invited, was to review the cost benefits of these two approaches. After almost two hours of detailed and tedious cost analysis, John Naugle, the head of the space science and applications department at headquarters (having taken over that position when Homer Newell was promoted to associate administrator of all of NASA in 1967), recommended that NASA go with Marshall's associate contractor approach.

As the meeting was winding down, Fletcher suddenly changed the subject entirely, using the last fifteen minutes to force a rethinking of the very nature of the space telescope project. As George Levin noted in his minutes of the meeting, "Fletcher went on to redesign the LST."[6]

To understand the context of Fletcher's actions, as well as much of the history of the Hubble Space Telescope, it is necessary to also understand the annual and cyclical nature of NASA's budget negotiations. These negotiations last two years: one year for NASA and the executive branch to work up their proposed numbers, and one year for Congress to review these numbers and to approve or change them.

Take the 1974 NASA budget as an example. The negotiations for this budget began inside NASA headquarters in 1972, with NASA headquarters bureaucrats meeting repeatedly with NASA center bureaucrats and White House bureaucrats to argue about how to divide up the money. Then, at the beginning of 1973, the 1974 budget was finalized and presented with great fanfare to Congress. Congress then reviewed and debated this budget for the better part of 1973, holding hearings, writing reports, and arguing its substance on the floor of the Senate and House before finally giving its approval.

Each year, this cycle is repeated, and each year, it influences how projects live or die.

Because the space telescope was not yet an official NASA project, its place in the 1974 budget was not significant. Yet, everyone knew that once it got under way it was going to quickly become one of NASA's most expensive projects. Thus, Fletcher's main concerns during that December 1972 meeting centered not so much on how many contractors NASA hired to build the space telescope but on the future 1974 congressional budget negotiations and how an expensive LST project was going to affect them. As he listened to his managers debate the pros and cons of these two contractor approaches, Fletcher was startled by the total proposed cost numbers. Goddard's proposal estimated a price for LST through launch and one year of operation at $500 million. Marshall in turn figured the telescope would cost about $900 million.

Fletcher waited until the main business of the meeting was finished before raising his concerns. He pointed out that it was unrealistic to plan on building a billion-dollar program, considering the political atmosphere in Congress at the time. He insisted instead on a budget limit of $300 million for construction, launch, and the first year of operation. He also demanded that the LST be simplified, reducing the number of science instruments as well as the number of shuttle flights to maintain it.

Immediately afterward and for years to follow, Fletcher's magical $300 million number, picked out of the air with no connection to actual cost, was the bottom line for the Large Space Telescope. As headquarters bureaucrats went ahead with Marshall's two-contractor arrangement and began actual construction, they did so using Fletcher's number, limiting the cost through launch to between $290 and $340 million.

What they did not do, however, was simplify the project as Fletcher had stipulated. By the end of 1972 the basic audacious design for the Large Space Telescope was essentially complete. No one wanted to change it significantly. For the astronomers, a significantly smaller telescope wouldn't be worth building. For the engineers, a simpler telescope wouldn't be as challenging to build.

Thus, not only would the Large Space Telescope have a mirror 120 inches across, making it for the time the third largest telescope ever made, NASA was going to put it in orbit, using the shuttle and astronauts to maintain and update it periodically, as recommended by Spitzer's astronomy committee at the 1965 Woods Hole meeting. All the design studies, by Goddard, Marshall, Boeing, and others, had conceived the space telescope for a time when space travel was routine, ambitious, and continuous, with the space shuttle making anywhere from 25 to 60 flights per year. It had therefore been planned as an automated and permanent observatory, operated remotely from the ground but with the capability to add, remove, and change out instruments, just as is done with any telescope on the ground.

The heart of the telescope would be its mirror and structure, but behind that mirror the telescope would have the capability of taking on any type of research instrument, from spectrographs to optical cameras to infrared or ultraviolet detectors. And as with all ground-based telescopes, as new developments made old instruments obsolete, the instruments in the telescope could be changed out as well. Moreover, the designers saw that the most efficient way to do these manned maintenance missions was with unscheduled shuttle flights. Rather than plan specific regularly scheduled upgrades, they expected shuttle missions to be so routine that whenever the telescope needed repair or upgrade, a flight could be quickly scheduled and inserted into the shuttle program.

As absurd as it seems, accepting Fletcher's lowball budget number without changing any of the telescope's design was (and remains) the normal way of doing business in Washington. Called by Beltway insiders a "buy-in," this Machiavellian approach to government funding relied on the fact that once the project was under way it would be difficult for Congress to kill it. Instead, the bureaucrats and their advocates outside the government would get Congress to approve the project at a lower number, then come back later for more money when needed.

Even as these bureaucratic machinations were taking place at NASA headquarters, O'Dell was starting development work on the space telescope's science instruments. In December 1972, he put out a call to the scientific community, asking them to propose instrument concepts. In conjunction with this call O'Dell also organized what he called a "dog and pony show," traveling the world in January 1973 to encourage astronomers to get involved. "The show opened at Harvard, right there at Harvard College Observatory, right in their library," O'Dell remembered. "Then we went to Chicago. Then we went to Caltech."[7] Eventually, O'Dell even went to Europe to press the flesh with scientists and make the case for the Large Space Telescope.

By April O'Dell was able to sift through the various concept proposals and form a number of design teams, all of which became part of what he called the Large Space Telescope Working Group. Eventually made up of seven teams—one for each of the telescope's six instruments plus a group of "interdisciplinary scientists" whose job was to take a larger view of the project—the initial working group met four times during the last half of 1973 as it determined the design of the telescope's science instruments.

By late 1973 when Naugle submitted his 1975 science budget, the project had advanced enough that he listed the telescope as a separate item, $6.2 million to fund the continuing design work. No one objected, and the item was included in the budget proposal that was presented to Congress on February 4, 1974.

Throughout the early part of 1974 the budget floated through Congress with little comment and no objection. During the February hearings in front of both the Senate Aeronautical and Space Sciences Committee and the House Science and Astronautics Committee, there were few fireworks, with most of the questions centering on the just completed Skylab program and the ongoing construction of the space shuttle fleet. Though there were some skeptics of NASA, the majority of the elected officials on these committees were very supportive of NASA. As Senator Frank Moss (D-Utah) noted at the opening of the Senate hearings, "I am pleased to see the NASA budget request for next year at a higher dollar level, albeit only slightly higher."[8]

Meanwhile, O'Dell found that the bulk of his in-NASA budget discussions with Naugle centered on whether to officially introduce the telescope as a major new project with full funding in next year's 1976 budget, or wait one more year.

Things changed, however, when Fletcher and Naugle gave testimony before the House Subcommittee on Appropriations in late March. Here they found that the winds of change were blowing hard, in their face.

NASA's budget for 1975 was part of an omnibus bill that also included funds for the Veterans Administration, the Department of Housing and Urban Development, the National Science Foundation, the Federal Communications Commission, and a handful of other smaller agencies. With the last U.S. troops only recently out of Vietnam, the congressmen on the Appropriations Subcommittee that handled these agencies were far more interested in getting money to help disabled veterans than they were for funding a new NASA project. Moreover, the increasing U.S. dependence on foreign oil made energy programs significantly more appealing to them than space, since "people are starting to think about more earthly problems," noted subcommittee chairman Edward Boland (D-Massachusetts). "We have a pretty good problem on Earth with respect to energy, and so there has been some concern about the money we are spending in space, and whether we ought to continue that."[9]

This subcommittee then devoted three days of hearings to painstakingly dissecting the NASA budget, going over every detail and budget item, searching for things to cut as well as ways to make NASA shift the focus of its research toward Earth-based environment work, such as energy conservation, solar power, or even hydrogen-powered cars.

When the subject of the Large Space Telescope came up on day one, Naugle was asked how much the project, once started, would eventually cost. At this point he made his first tactical error, stating the cost would be "$400 to $500 million," based upon the full fifteen-year mission, instead of Fletcher's $300 million figure through the first year of operations. As Fletcher had expected, this number was immediately noticed by the congressmen. Fletcher promptly tried to deemphasize it by adding that Naugle's figure was to "be phased over quite a number of years."[10]

Things got worse on the second day of hearings. In the morning session, congressmen George Shipley (D-Illinois) and Burt Talcott (R-California) expressed strong reluctance to fund more space telescope studies. Both noted how the National Science Foundation had recently asked for and gotten funding to build the Very Large Array radio telescope in New Mexico. As Talcott said, "We are exploding with telescopes."

Shipley also worried about the possibility of a buy-in. "The thing that bothers me again is that this is more or less a new program. You are talking about a possible estimate of $400 million." He then added, "It would be very unusual if you asked for the $6.2 million for a definition study and did not come in next year for development funds."

Naugle then made his second tactical error, noting how a major review by the National Academy of Sciences of the future of astronomy had been completed in April 1972. Chaired by Dr. Jesse Greenstein from the California Institute of Technology, the report, dubbed *Astronomy and Astrophysics for the 1970s*, had been the result of four years of discussions among almost all the important astronomers in the country. The report was also the first of what was to become regular decadal planning surveys by astronomers of their future priorities. In describing this report, Naugle explained how it had "looked at all of astronomy, ground-based and space-based, and [had] recommended an order of priority." He then added, "We have been following the guidelines in that report. We moved through the sequence that they recommended. The Large Space Telescope was recommended as an effort we should do."[11]

What Naugle didn't mention, however, was that though the Greenstein report had listed LST as a priority, it had given it a much lower priority than a lot of other projects, placing it ninth on a list of eleven.[12] Having now made the congressmen aware of the report, Naugle should have known that they would immediately get a staffer to find and review it.

By the afternoon session of the second day, support for the space telescope continued to deteriorate. As the congressmen made their way point by point through the NASA budget, they discovered that the agency was also planning to build a ground-based 120-inch infrared telescope on the top of Mauna Kea in Hawaii for $6 million. Because ground-based telescopes were normally funded out of the National Science Foundation, Boland immediately asked, "Why do you have this in your budget? Why should not this be funded by the National Science Foundation?"

Fletcher tried to explain, saying that the infrared telescope was being built to support NASA's planetary missions.

At this point Congressman Robert Tiernan (D-Rhode Island) brought up the Greenstein report again. Having now reviewed it, he pointed out that the infrared telescope was listed third among the report's four top priorities, none of which included an optical space telescope. Naugle was then forced to agree to provide for the record the Greenstein report's full recommendations, thereby making sure that everyone on the subcommittee would read it.[13]

Thus, when the Appropriations Subcommittee submitted its final report to Congress on June 21, it wasn't surprising that they recommended cutting *all* funding for the space telescope. As their report noted, "The LST is not among the top four priority telescope projects selected by the National Academy of Sciences, and suggests that a less expensive and less ambitious project be considered as a possible alternative." Moreover, the subcommittee also recommended a total of $39 million of cuts in NASA, while also including a comparable $39 million increase for the Veterans Administration. These cost-cutting congressmen, focused on spending money on "earthly problems," had found with the Greenstein report the justification they needed for cutting both NASA and the space telescope.[14]

In the House debate five days later, these changes met with general approval. If anything, there was concern from many members of Congress that the increases for veterans were insufficient. Though a few representatives questioned the NASA cuts, including James Symington (D-Missouri), who made a speech decrying the deletion of LST and noting that Jesse Greenstein actually supported the telescope, the budget passed by 407 to 7. Even Symington voted for it.[15]

Once again, despite years of effort, Lyman Spitzer's dream seemed about to evaporate.

In Huntsville, Bob O'Dell wasted little time. He had heard about the House report on June 20, the day before it was released, and had immediately started a campaign to get the money restored. O'Dell quickly recognized that the main justification used by the subcommittee for rejecting the space telescope was the low priority given to it by the *Astronomy and Astrophysics for the 1970s* report. He surveyed the report's members, and discovered that of its twenty-three members, fourteen were already working on the space telescope in some capacity.[16] It was obvious that he had to get these people to tell Congress that they considered the space telescope more important than indicated by their report.

For O'Dell the situation was delicate. His three-year leave of absence from his tenured position at the University of Chicago would end in 1974. He didn't want the project to die, but he also needed to know if there would be a project at all in order to decide whether to hold on to his position in Chicago. More importantly, as a civil servant he was legally forbidden from lobbying Congress. "I couldn't call, but I could pick up the phone," O'Dell explained.[17] And he could work to get others to do his work for him.

O'Dell began contacting every important astronomer he could think of for help. In the five days between the release of the House subcommittee report and the House vote, O'Dell worked the phone aggressively.

He did this despite being specifically instructed by his administrative boss at NASA headquarters, Bland Norris, to do nothing.[18]

First on his list was Jesse Greenstein. Of all the members of the 1972 report, Greenstein—the report's chairman and the man who had organized and edited it—was known to have decidedly ambivalent feelings toward space astronomy, beginning at the very dawn of the space age. In the late 1940s the Army was learning how to build and launch missiles by flying the German V2 rockets that they had captured at the end of World War II. Rather than fly them empty, however, they offered the bomb chambers to scientists, letting them fly scientific experiments at almost no cost. The result of this collaboration had produced Richard Tousey's first ultraviolet spectrums of the Sun as well as James van Allen's first detailed mappings of the Earth's magnetosphere.

Greenstein, then a thirty-eight-year-old astronomer working at the Yerkes Observatory, decided to give it a try. He built and installed a spectrograph on one V2 rocket, and watched it blast off from the White Sands test range.

All he got from the experiment was blank film. The shutter of the spectrograph never opened.[19]

This failure turned him off from space astronomy, and made him for decades one of the most steadfast opponents to the idea of spending large sums of money for the building of space telescopes. Though he repeatedly said that he did not specifically oppose all spending on space, he always treated any space proposal with great skepticism. As he once told Lyman Spitzer after Spitzer had started work in 1961 on the Copernicus satellite, "Lyman, you're young, you'll live to see it fail."[20]

In chairing the 1972 report, however, Greenstein tried very hard not to impose his personal biases on the report. "I don't make this report. I'm chairman," Greenstein explained in 1978. "And I didn't try to influence its outcome."[21]

As a result, the survey's first four recommendations included "a program for x-ray and gamma-ray astronomy from a series of large orbiting High Energy Astronomical Observatories." The report's number one priority, however, was the building of a very large array radio telescope, a recommendation that Congress had accepted when it approved $80 million of funding in 1972 for the construction of the Very Large Array. And though the report advocated building the space telescope, it gave its construction a priority below things like orbiting solar observatories and the use of balloons, airplanes, and suborbital rockets.[22]

For Greenstein, the report's emphasis on space astronomy was less distressing than its focus on funding big government science projects rather than giving money to university research. "[The report] ended up with all my business claims for a balanced program for astronomy disappearing down the maw of big science; the death of university astronomy."[23]

When O'Dell called Greenstein about Congress's rejection of the space telescope, Greenstein was surprisingly agreeable about helping. He explained that the reason the survey had placed the space telescope so low in its priorities is that NASA officials had told them to consider the telescope a shuttle-linked project which would be built in the 1980s, not the 1970s. As Greenstein noted later, "We couldn't tell them to put up the space telescope when there wouldn't be shuttles in the '70s to fly it and repair it."[24]

About this same time O'Dell also received a call from the office of Congressman James Symington (D-Missouri). Symington, a member of the House Science and Astronautics Subcommittee that only weeks before had approved NASA's request to do the telescope studies, was upset by the House Appropriations Subcommittee report. He considered the deletion of LST's funds an invasion of his turf, and intended to fight it.

In talking to Symington's office, O'Dell learned that Symington wanted to get Greenstein to clarify the position of the National Academy report, to explain why LST had been given such a low priority. O'Dell, who knew that Greenstein was attending a meeting in DC, provided them hotel contact information and suggested they arrange a meeting with Greenstein. As a result of this meeting Symington was able to cite Greenstein's name in his House speech on June 26, announcing further that Greenstein was writing a letter in support of the space telescope, and that Symington intended to read that letter into the Congressional Record as soon as he got it.

■ ■ ■

Even as O'Dell was working the phones in Huntsville, in Princeton Spitzer and others were gearing up their own campaign to save the project. The most important person working with Spitzer was a man who just three years earlier had not even heard of the space telescope.

When O'Dell's "dog and pony show" had toured the country in January 1972 to sell the space telescope to astronomers, the buzz he created had reached the ears of John Bahcall at the Institute for Advanced Study in Princeton, New Jersey. Bahcall, a theoretical physicist whose models describing the nuclear processes within the Sun had revolutionized the field in the 1960s, had only recently become interested in the study of more distant astronomical phenomena.

Bahcall had just gotten his appointment to the institute in 1971 and was therefore ideally placed to become a crucial member of the space telescope team. For one thing, the institute—where Einstein had worked for the last twenty-two years of his life—had been conceived and designed to provide an environment where brilliant minds were left free to follow whatever research goal appealed to them. As noted by journalist Ed Regis in his book *Who Got Einstein's Office?*:

> The institute is exactly the nirvana for eggheads that it's cracked up to be. The temporary members, and the far smaller group of permanent professors—there are about twenty-five of these—work on their own projects at their own pace and have no further responsibility or accountability to anyone. . . . You don't even have to write up a report describing the work you did there, if any.

Not only did the institute pay its professors a very high salary, it provided them an office, an apartment, and breakfast and lunch five days a week as well as dinner on Wednesdays and Fridays.[25]

Bahcall thus did not have to teach, and was given no research assignments by anyone. He could follow whatever interests appealed to him.

Moreover, the institute was located walking distance from Princeton University, where Lyman Spitzer headed the astrophysical department. When Bahcall arrived he took the very informal weekly luncheons that Spitzer and Schwarzschild had been holding since the fifties and transformed them into a major social and academic event, occurring every Tuesday in a board room at the institute and attracting as many as a hundred people from all of the nearby universities. Rather than have people simply sit and chat with their neighbors, Bahcall would instead ask a number of people to describe either their research or some new

Fig. 3.2. John, Neta, and Dan Bahcall shortly after Dan's birth, 1971. (Photo courtesy of Neta Bahcall)

exciting development in their field. "Any visitors would always be called upon," remembered Neta Bahcall, astronomer as well as Bahcall's wife. "It was a wonderful opportunity—that's why it became so popular—for all of us to hear about everything that is happening in astronomy."

As the organizer of this semiformal event, Bahcall added his own personal touch, heralding each luncheon's start by standing in the stairwell just outside his office and shouting at the top of his lungs that the lunch was about to begin. "He liked to have fun," Neta Bahcall said. "We would just jump!"[26]

Surrounded by such a vibrant intellectual community, Bahcall—who was naturally enthused about the pursuit of knowledge anyway—couldn't help but be exposed to a whole range of new topics. As soon as he heard about the space telescope he was inspired by the idea, and immediately sent off a short letter to O'Dell to express his excitement and offer his assistance should it be needed.

When O'Dell got this letter, which was by no means a formal proposal and wasn't intended as such, he instantly decided that that was what it was. "John is one of those guys with just endless enthusiasm and energy," O'Dell said in a 1985 interview. "If you're around him for too long, he's too much, because sometime you just desire quiet and sleep and rest."[27] He called Bahcall and asked him if he would like to become part of the science working group that O'Dell was forming. O'Dell wanted Bahcall to be one of the space telescope's interdisciplinary scientists, along with Spitzer, Art Code, Margaret Burbidge, and a handful of other prominent astronomers.[28]

Though Bahcall hadn't planned on getting that involved, the idea of helping design the world's first space telescope was too alluring. "It was just the greatness of the program, the potentialities of the program, independent of whether I was a theorist or an observer, which attracted me."[29] He agreed to join O'Dell's science working group, and for the next year or so was involved in the typically unexciting and complex legwork necessary to finalize the telescope's basic scientific design.

When the House subcommittee declared the project dead on June 21, however, John Bahcall's participation was suddenly pushed to a new level. He became O'Dell's lobbying voice, saying the things to members of Congress that O'Dell was forbidden to say because of legal restrictions. "No one had more conviction of the rightness and the value of [the] space telescope," noted O'Dell. "I was feeding him ammunition and he was just blasting."[30]

John Bahcall was a soft-spoken but firmly clearheaded and determined man. His journey to the world of astrophysics was strange and truly American in nature. Born and raised in a lower-class Jewish family living in Shreveport, Louisiana, he had little scientific training as he grew up. "My family and my family's friends were not intellectuals. They concentrated on making a living, a fulltime occupation."

In high school his interests were more focused on playing tennis then learning about science: his school day ended at noon rather than 3 pm so that he could practice for the school's tennis team. Only in his senior year did he discover that he had a talent for intellectual pursuits. That year he joined the debating team and, to his glee, did so well that he quickly moved up to the school's first team. "When John set his mind to doing something he pretty much always did it," said Max Nathan, his school friend and debating partner. The two went on to win both the state and regional competitions, making them the first team from Louisiana eligible to go to the national tournament, held in Boston in 1952. They traveled to Boston on their own and beat students from some of the best schools in the country to become the national champions.

Bahcall then went to Louisiana State University, studying philosophy with the thought that he would use it toward becoming a reform rabbi. During the summer break, however, his mother and a cousin arranged for him to transfer to the University of California at Berkeley. Though he was still focused on philosophy, Berkeley required that he take at least one science course. Up until then Bahcall had studied no science, and in high school he had not progressed past elementary algebra.

His advisor told him that he would have to make up this lack by taking evening high school science classes. Bahcall, however, didn't want to go back to high school. He looked over the various available physics courses. There were three choices, one for non-scientists, one for engineers and medical students, and one for those who planned to become professional physicists. Bahcall went to the teacher of the last option and said he wanted to try it. Despite his lack of prerequisites, Bahcall was allowed to enroll, on the condition he drop the course as soon as he realized the work was beyond him.

Instead, he persevered, pulling a C in the course. "It was the most difficult thing I had ever done in my life," Bahcall remembered in 2002. It was also the most exhilarating thing he had ever done. "I fell in love with the subject."[31]

In the ensuing years Bahcall focused his interests on two subjects, the physics behind the Sun's nuclear processes and the nature and makeup of the just-discovered quasi-stellar objects, or quasars. By the 1960s he was involved in the first neutrino telescope, being built by Ray Davis of Brookhaven in the Homestake gold mine in South Dakota. Davis needed to know if his telescope would be sensitive enough to detect the predicted flux of neutrinos that the Sun produced as a by-product of its nuclear engine. Bahcall worked out the first detailed analysis of that process, predicting that the flux would be large enough for Davis's telescope to measure it.

Bahcall was only partly right. Though Davis's telescope was successful in detecting the neutrino flux, to everyone's puzzlement the flux was significantly less than predicted by Bahcall's model. It wasn't until the 1990s that scientists realized that the difference was explainable only in terms of new physics. Neutrinos were more complicated than expected: they had mass, and would oscillate between different states as they traveled through space.[32]

In 1971 Bahcall accepted a position at the Institute for Advanced Study in Princeton. There he planned to dedicate his time to studying quasars and solar neutrinos while also strengthening astrophysical research at the institute. As Ed Regis wrote in 1987:

> When [Bahcall] became executive officer of the School of Natural Sciences [at the institute], he was not at all bashful about packing the place with astronomers. These days, when institute administrators have to scramble to find office space, or room in the housing project, for visiting members, they blame it on John Bahcall, who brings in astronomers by the dozens. The institute's astrophysicists, coupled with the university astronomers next door, have made Princeton a world center for astrophysics.[33]

Within hours of learning of the 1974 congressional vote to delete all space telescope funding, however, Bahcall and Spitzer found themselves to be lobbyists instead of scientists. As a professor, the timing was excellent for Spitzer. He had all of July to devote to the campaign. Bahcall's time was his own. His only loss—which was not trivial—was the time the lobbying took him away from his research.

While Spitzer started calling his contacts in Congress, which were many because of his work on nuclear fusion, Bahcall took out the directory for the American Astronomical Society and began calling everyone he knew to ask them to write letters to their senators and representatives.[34]

Spitzer also called O'Dell, and the two brainstormed what to do. They decided that while Spitzer and Bahcall contacted astronomers, O'Dell would call the many industry contractors working on the space telescope and get them to funnel information to Bahcall and Spitzer as well as to Congress.

First on their agenda, however, was to clarify the reasoning behind the Greenstein report's recommendation. Working together, Bahcall, Spitzer, and O'Dell solicited all twenty-three members of the Greenstein committee and got them to endorse a draft statement by Bahcall that stated, "In our view, Large Space Telescope has the leading priority among future space astronomy instruments."[35]

Then, on July 2, Congressman Symington stood on the House floor and, as he had promised on June 26, inserted into the Congressional Record a letter by Jesse Greenstein in support of the Large Space Telescope. As Greenstein wrote, "Astronomers felt then and feel now that the LST is the ultimate optical telescope and that together with a balanced ground-based program, it will open up new vistas for the human mind to contemplate."[36]

By that date Bahcall and Spitzer had already arranged a series of meetings in Washington, DC, through July, visiting with numerous elected officials and their staffs, including senators William Proxmire (D-Wisconsin) and Charles Mathias (R-Maryland). These two men were respectively the Democratic chairman and the senior Republican on the Senate Appropriations Committee that was to review the House budget bill and make its own budget recommendations to the Senate.[37]

Proxmire, well known for his Golden Fleece awards for what he believed was absurd and wasteful government spending, was of like mind with Boland in opposing spending at NASA. No one expected to get much help from him. Mathias, however, was very interested in helping. After their meetings he agreed to reinsert the space telescope's $6.2 million in the budget.

For Mathias, a liberal Republican whose willingness to oppose the Nixon administration had irritated the leadership of his party, the space telescope was a chance to show his broad-mindedness to Maryland voters. He expected a tough reelection campaign later that year, first in the primary from Republican conservatives, and then in the main election from Barbara Mikulski, a Democratic city council member in Baltimore.*[38]

For Bahcall, the lobbying experience that summer was somewhat frustrating. "It came at a large cost to me personally, in terms of not allowing

* In the end, Mathias beat Mikulski handily, 57–43 percent. She in turn would return to win the 1986 Senate campaign when Mathias retired.

me to do science during that period."[39] At times it was also nerve-racking. In one case, Bahcall had wrangled an appointment with Congressman Max Baucus (D-Montana), also a member of Boland's Appropriations Subcommittee. Bahcall had heard that Baucus had a lawyer in his office who was married to someone who worked at Goddard. Bahcall called the wife, who put him in touch with the lawyer, who arranged the interview.

Bahcall was ushered into Baucus's office, put on a stool, and told that he had 15 minutes to make his pitch. Worse, Baucus began firing questions at Bahcall, one after another. "I had the feeling that . . . if there were any questions I couldn't answer in 15 seconds, he was just going to walk out of the room in disgust."[40]

At times the effort seemed pointless. For example, Bahcall and Spitzer spent fifteen minutes trying to explain the importance of the project to Democratic congressman George Shipley. Shipley, one of the telescope's biggest skeptics in the House Appropriations Subcommittee, explained that only two years previously Congress had approved another big-budget telescope, the Very Large Array (VLA) radio telescope to be built in New Mexico. "We just approved the VLA telescope, which was going to solve the riddles of the universe," Shipley complained. "Now you want another telescope. What's wrong with you fellows?"[41]

Shipley's complaint was typical. The Greenstein decadal survey's recommendation, with VLA at the top and the space telescope down near the bottom, had given congressmen like Shipley the clear impression that the astronomers themselves believed that the space telescope could wait.

In spite of all this, the lobbying effort by Spitzer and Bahcall bore fruit. When the Senate Appropriations Committee issued its report on August 1, it not only included the money for the LST, as Mathias had promised, it also reinstated the $39 million that the House had tried to cut from NASA. The Senate was far less cost conscious than the House. They not only accepted NASA's budget numbers, they also topped the House's increased appropriations for the Veterans Administration by more than half a billion dollars.

Proxmire, who opposed these budget increases, wasn't about to give up. In bringing the budget bill to the Senate floor on August 5, he introduced an amendment to cut an additional three percent across the board from the discretionary portions of the budget bill. In the floor fight that followed, Proxmire's amendment failed. Not only were most senators uninterested in cutting spending, they were terrified that someone might think they were in favor of reducing the budget for the Veterans Administration.

Proxmire's maneuver, did, however, succeed in getting the budget bill sent back to the Senate Appropriations Committee for review. As he announced just before the Senate made this decision, "I intend to fight and fight to the very best of my ability for at least a $160 million deduction."[42] Though the subsequent second report, issued on August 15, still included an increase for the Veterans Administration exceeding half a billion dollars, the increase was now offset by other trims. For example, even though the NASA budget still included the $6.2 million for the space telescope, the agency's budget was only increased by a total of just under $4 million. The committee had merely left the decision on what to cut to NASA. "The Committee feels that NASA is better able to apply this reduction with a minimum of disruption to its priorities."[43]

On August 16 the final conference committee negotiations between the House and Senate took place. Though Mathias was on the committee, so were Boland, Talcott, Shipley from the House, and Proxmire from the Senate, making it amazing that many of the Senate's higher figures were accepted. For NASA, however, the money for telescope studies was cut again to $3 million, a trim that meant that though work could go forward, it would be impossible to introduce the telescope as a new project in 1975. The project would have to wait at least one more year.

Moreover, the cut indicated that the project still faced strong opposition. On August 2, even as negotiations were ongoing, Spitzer called O'Dell to tell him of a recent conversation he had had with James Fletcher, NASA's administrator, in which the idea of reducing the size of the telescope had come up. As then planned, the telescope's mirror had a diameter of 3 meters, or about 120 inches. Because Spitzer had the impression that Fletcher's support of the space telescope was somewhat lukewarm, he thought it wise for the project to consider Fletcher's suggestion. O'Dell agreed. They had to do something to improve their support in both Congress and NASA. O'Dell suggested that Spitzer write him a letter outlining his thoughts on the matter. On August 7 Spitzer did so, suggesting that the science working group "consider telescopes with an aperture ranging from 1 to 3 meters . . . The chief consideration should be given to an alternative [large space telescope] with an aperture between 2 and 2.5 meters."[44]

From O'Dell's perspective, 1.8 meters "would have been wonderful," though he believed that they could get the best compromise of good science and greatest savings by going to some middle size, say 2.4 meters. Moreover, he was aware, because of his top secret clearance, that the military space program had already built and launched high-resolution spy satellites using mirrors with a 2.4-meter diameter. Using this diameter in the space telescope would save money, since the development costs for a larger mirror would be far greater. Over the next six months he made an effort to poll every scientist involved in the project on this proposal. Most were also willing to consider reducing the mirror size, but rejected anything less than 2 meters. Even at that diameter John Bahcall told O'Dell that he would have to reassess his personal role in the project.[45]

In the end, the astronomers compromised, and at a December 13, 1974, working group meeting at Marshall they agreed to a reduction from 3 to 2.4 meters, or 120 to 94 inches. This smaller size would still allow them to achieve most of their scientific goals, while simplifying construction enormously. Though a 3-meter telescope could fit inside the shuttle cargo bay, the fit would be very tight, and carried with it a lot of difficult engineering problems that no one was sure could be solved easily or cheaply. The smaller size made these engineering problems much more manageable.[46]

After these 1974 compromises O'Dell and the other astronomers assumed that the telescope would finally be submitted for full funding in the 1977 fiscal budget and submitted to Congress in January 1976. Called a "new start" by NASA budget managers, the project's entire costs would now be listed in detail in the NASA budget.

The 1974 budget battles, however, indicated the tenuous nature of the project's future. The attitude of the Congress in the late 1970s was far from enthusiastic about space and NASA, and after the November 1974 elections, the political situation had become even worse. The Watergate scandal and Richard Nixon's resignation in August 1974 had resulted in a new Congress dominated by Democrats like Edward Boland, individuals who were much more inclined to spend government money on social programs, not big scientific research projects involving astronauts and NASA.

For the '77 budget the agency had been planning to include two science "new starts," the space telescope and the Solar Maximum Mission, intended to study the Sun during its solar maximum in 1979–80. In 1974 John Naugle had retired, and the new associate administrator for space science, Noel Hinners, realized that he couldn't justify a new start for the space telescope in '77. Not only was the new Congress more hostile to new NASA spending, the new president, Gerald Ford, wanted to cut the federal budget as well. In October 1975, Ford had announced that he wanted to trim an additional $28 billion from the total 1977 budget, of which NASA had to shoulder $305 million in cuts. In order to meet these demands, Fletcher, still a tepid supporter of the space telescope anyway, realized that he did not have the cash to start the telescope in 1977, even at the $300 to $350 million cap he was demanding for it. Moreover, while the Solar Maximum Mission had to be in orbit in time for the solar maximum in 1980, the telescope could wait. As several congressmen had previously noted, the stars would always be there. Fletcher told Hinners to hold off.[47]

Normally, a bureaucrat in Hinners's position would have simply included a few million dollars so that the telescope's design work could go forward, pending a new start some time in the future.

Hinners, however, had a different idea. "I figured that if there were one or two million in there, that would probably keep the outside community quiet [but] wouldn't have done much." He decided to exercise what he liked to call "the Black Art." Instead of simply accepting the delay and including a small stipend to keep the program alive, he deliberately deleted the entire telescope budget. "My strategy was to put nothing in and get the outside world really antagonized. And it worked. Panic!"[48]

As part of the annual cycle of budget negotiations, the President's proposed federal budget is kept under wraps until mid- to late January. No one outside the executive branch is supposed know what's in it until it is revealed by the White House. Nonetheless, on December 22, 1975—three days before Christmas—O'Dell called John Bahcall and leaked the fact that the telescope had been deleted, thereby giving Bahcall about a month to prepare his lobbying campaign before the official release of the budget on January 21, 1976.[49]

At the time Spitzer was on sabbatical in Paris, so it was entirely up to Bahcall to get the campaign underway. Once again he started calling and writing everyone he knew, suggesting that they draft letters to be sent the instant the budget became public knowledge. He himself drafted a letter to Fletcher, expressing his "surprise and disappointment," and his fear "that the scientific community will conclude that NASA is intent on obtaining funds to support its own institutional needs in preference over wider goals." He also got Spitzer and several other important astronomers to agree to sign a joint statement to Fletcher in which "all signatures were made by John Bahcall with the agreement of the signees so that [Fletcher] could receive this letter as soon as possible." This joint letter described "the immediate angry reaction of many scientists" to the deletion and demanded a meeting with Fletcher to discuss the issue.[50]

Unlike the 1974 lobbying effort, the 1976 effort carried risks. In 1974 the campaign had the support of the bureaucrats in NASA. In 1976 the campaign was directed *at* the bureaucrats at NASA, criticizing their decision to delete the telescope. If it should fail, the astronomers might only succeed in antagonizing the very people they had to work with at the space agency.

In this context it is unclear how much Bahcall knew about Hinners's strategy. What is known is that he did talk numerous times to Hinners immediately after finding out about the cut, working out the campaign strategy with him. Moreover, as Bahcall himself said many years later, there were many individuals at NASA, such as O'Dell, who "were willing to do courageous and illegal things" to help save the telescope. For example, O'Dell was pushing the limits of the law if not breaking it by his behind-the-scenes lobbying effort.[51]

For O'Dell, the risks were significant. Following the 1974 lobbying campaign he had written the University of Chicago to formally give up his tenured position there. As he explained in his resignation letter, "My firm belief is that the Large Space Telescope should, and eventually will, be built. My departure now would be interpreted as a loss of support for the program. . . . In light of the current difficulties, this interpretation would clearly be enough to kill this program. I cannot in good faith do this and must, therefore, tell you that I do not intend to return to the University of Chicago next autumn."[52]

Then, only a few weeks before Hinners deleted all money for the space telescope, George Pieper at Goddard offered O'Dell the job of running the center's Optical Astronomy department. Though the two men disagreed strongly on many issues relating to the space telescope, they also respected each other. The job would give O'Dell a more secure position in space science, especially considering the space telescope's lack of funding. After taking several weeks to think about Pieper's offer, O'Dell turned it down, despite having learned in the interim of the space telescope's loss of funding. "The relocation of my family three years ago had long term effects on my children which I only now really appreciate," he explained in a letter to Pieper. Moreover, his wife liked it in Huntsville, and did not want to move.

O'Dell was now totally committed to the project. "Like Cortez burning his boats on the beach, I was going to make sure [the space telescope] succeeded." Without tenure, without funding, and without any clear future at NASA, he was left with only one choice: succeed.[53]

As soon as Fletcher officially announced NASA's budget in mid-January, the campaign began, with Bahcall acting as public point man and O'Dell providing him and others information and support. Once again there were trips to Washington, DC. Once again there were endless phone calls and letters. Once again Bahcall found himself meeting with senators and congressional representatives.

Though his 1976 trips to Washington with Spitzer and others to lobby elected officials were important, Bahcall's greatest influence in 1976 was as the campaign's lead organizer. He began issuing regular reports to the astronomy community, providing specific details about the various members of Congress on the important budget committees, including summaries of their attitude toward the space telescope and what might influence them. Under his leadership a legion of other scientists took up the banner of the space telescope and carried it wherever they went.[54]

For example, inspired by Bahcall, George Wallerstein of the University of Washington made his own lobbying visits to Congress, sitting down with the staff of two representatives and one senator. He then sent out a letter addressed to "numerous astronomers," haranguing them to write their own letters. "Do it TODAY. If you don't, I don't want to hear one peep out of you complaining about Congress or the lack of jobs for your students in the 1980's; it will be your fault, not theirs!"[55]

Also inspired into action by Bahcall was George Field, director of the Harvard College Observatory and the Smithsonian Astrophysical Observatory and chairman of the Physical Sciences Advisory Committee that regularly provided advice to Fletcher as well as the White House. Field had been a student of Lyman Spitzer's, and had adopted the same research interests as Spitzer after his graduation, investigating the gases lurking out in the empty regions between the stars. In fact, he had specifically picked Princeton for his postgraduate education because of his admiration of Spitzer and Schwarzschild. "I learned a tremendous amount from [them], both about the specifics of astronomy, but also, the way to do science. I found them enormously stimulating people."[56]

Field, as a Massachusetts resident, had a particularly important position. Boland, as chairman of the House Appropriations Subcommittee and their chief opponent in the 1974 battles, represented Springfield, Massachusetts.

In 1976 Boland was no less of a problem. In budget hearings in February he noted that though the telescope had been dropped from the budget, his subcommittee was "under pressure from about every conceivable institution that is in astronomy in the United States" to get those funds reinstated. Nonetheless, he questioned the need for the telescope in lieu of other priorities. "What good is it going to do when you get it out there? What does it do? What does it find?" Despite attempts by Fletcher and Hinners to defend the telescope's value, Boland concluded by noting, "There are some problems here on Earth which we have to be concerned about, and they are the difficult ones. If we don't solve those, there won't be any money for space applications or space technology or space science."[57]

In his own effort to save the telescope, Field called Boland's office, wrote to him, and even talked directly to Boland at one point, all to no avail. His most interesting opportunity to influence Boland took place one day while he was at Washington's National Airport, waiting for a plane. There he saw the president of MIT, Jerome Wiesner, chatting with Democrat Thomas P. ("Tip") O'Neill, Speaker of the House, and the man who represented the congressional district in Massachusetts where Harvard was located. Field knew that Boland and O'Neill were old friends, having shared an apartment together in DC for the last twenty-four years.

Field joined them, and turned the conversation to the space telescope, explaining to O'Neill how the project might bring as much as $50 million in government money to Massachusetts. O'Neill found the thought of such pork very appealing, and promised to "discuss it with Eddie" when he and his wife joined Boland for a Cape Cod vacation in the next few weeks.[58]

O'Dell, meanwhile, continued to skirt the edges of the law in his effort to save the telescope. He would send out regular "Dear Colleague" letters to the members of the science working group, describing to them the status of the lobbying campaign and hinting at things they could do to further the campaign. He arranged for Noel Hinners to attend several science committee meetings run by O'Dell, including the space telescope's working group as well as a session of a science working group on doing astronomy on the space shuttle. At this second meeting one whole afternoon session ended up being focused not on shuttle astronomy but on "space telescope issues." While Hinners explained the logic behind the agency's budget decision (though not mentioning his own use of "the Black Art" to get things moving), O'Dell gently warned that if the telescope project did not get a new start by the next year, it faced the possibility of never flying at all.[59]

In the meantime, Bahcall, Spitzer, Field, and O'Dell continued to push for a meeting with Fletcher, which finally occurred, after several postponements, on May 19, 1976. To their horror, they discovered that Fletcher was not only relatively uninterested in the space telescope, he knew very little about it.

"We were appalled how little Fletcher understood [about] the project at the time," remembered Field in 1986.

"Fletcher himself did not come across as a person committed to science," remembered Bahcall in 1983. "He came across as a person . . . committed to the goals of NASA, which at the time were primarily institutional . . . and not particularly science-oriented."

Nonetheless, the meeting forced Fletcher to recognize how strong and deep the support for the space telescope was among astronomers. As Field noted, "Fletcher was a skilled bureaucrat." He now had to include in his political calculations this powerful and effective group of astronomers, a fact that made it far more difficult for him to oppose funding the space telescope.[60]

Despite all this campaigning, by early June the situation still appeared bleak. In his discussions with various senators and their staff, Bahcall was only able to get from them a small bone. Though no elected official was willing to reinsert the new start in the 1977 budget, a number of senators and representatives agreed to include language authorizing NASA to issue what was called a request-for-proposals (RFP). This document, which describes a project's specifications and is used by contractors to write their bids, is normally not released until after a new start is appropriated. Bahcall had gotten Congress to agree to allow the RFP's early release in anticipation of an official new start in the 1978 budget.[61]

On June 8, however, Boland stepped in to try and stop this maneuver. On that day his House Appropriations Subcommittee issued a report to the House in which they specifically prohibited NASA from releasing an early RFP or selecting a prime contractor. As the subcommittee report stated, "It felt that such a procedure established an unacceptable and highly unusual precedent."[62]

Boland's action was a serious tactical error. The chairman of the House Authorizations Committee, Don Fuqua, resented Boland's Appropriations Subcommittee interference in what he considered an Authorization Committee matter. At this same time it also appeared that Boland was getting sick and tired of the constant lobbying of astronomers. For example, in a meeting between astronomer Myron Smith and his own Texas congressman, J. J. Pickle, Smith was told by Pickle how "Boland was feeling quite harried over political lobbying by astronomers."[63]

In the House-Senate conference committee negotiations at the end of July, Boland's effort to stop the space telescope failed. The compromise went through. NASA was authorized to release its RFP early.

The battle was far from over. The compromise also stated that "this is not to be construed as an endorsement of development funding of the project."[64]

Over the next year the campaign went on unrelentingly. As one congressman noted, "[Many] of us have had our rugs and carpets worn out by astronomers." In February 1977, after NASA had released its 1978 budget with a new start for the space telescope, Bahcall was asked to testify before Congress. For this trip he also arranged numerous meetings with more members of Congress and their staffers, accompanied by astronomers George Wallerstein, George Field, Vera Rubin, and Margaret Burbidge, trying to stave off any opposition to the new start.[65]

Burbidge and Bahcall had reserved separate rooms in the same hotel, and had agreed to meet for breakfast before heading out to Congress. To Burbidge's surprise, Bahcall was five minutes late. "He's usually extremely punctual," she remembered in 1984.

As they ate, Bahcall explained that he had spent half the night in the emergency room. When he had been getting into bed the night before he had cut his thigh on a nail that had been sticking out of the mattress. "It was bleeding rather a lot," he explained. When he called the hotel desk they insisted he go to the hospital for stitches and a tetanus shot.

Anyone who has ever walked the halls of Congress knows the distances involved. Everything is spread out, with long tunnels between buildings. Yet, off they went, walking back and forth from office to office to see as many representatives as possible. "John was limping very slightly but going at full speed," Burbidge said in 1984. "He was obviously in pain."[66]

Once again, this lobbying effort had demanded an endless amount of networking. For example, getting to see Congresswoman Lindy Boggs, Democratic representative from Bahcall's hometown district in Louisiana and a new member of Boland's subcommittee, required a long chain of contacts. The journey began when Bahcall, Spitzer, and Burbidge met with Democratic senator Bennett Johnston of Louisiana. One of Boggs's staff members happened to be there, and in talking to this staff member Bahcall learned that Boggs's own daughter, Barbara Boggs Sigmund, happened to be a local New Jersey elected official in the county where Bahcall lived. Bahcall called Sigmund, and through her was able to arrange a meeting with Boggs.

The meeting itself was interesting. Because of an important pending House vote, Boggs arranged to meet Bahcall in the ladies lounge adjacent to the House floor so that she could get away quickly to vote if necessary. "For me the setting was somewhat bizarre," Bahcall remarked in a 1985 interview.

Boggs, however, was enthusiastic about the idea of a space telescope. Also, "it was somewhat helpful in Lindy Boggs's mind that I was from

Louisiana," Bahcall noted. She not only backed the telescope during March House Appropriations hearings, she helped Bahcall get appointments with a number of other important elected officials and their staff.[67]

Bahcall and Spitzer returned to Congress twice more in April 1977. Not only did they have to contend with Boland's opposition, but on May 11 Senator William Proxmire of Wisconsin announced his opposition as well, noting his lack of faith in NASA's cost estimates. As his press release noted, "A recent General Accounting Office report shows that NASA has neglected to report the full costs of the controversial space telescope project. According to the GAO, the space telescope will have a life cycle cost of $1.4 billion, or over 200 percent more than the NASA development estimate given to Congress."[68]

Though Proxmire was eventually proven right, his opposition was too little too late. The astronomers' campaign had worked. In mid-July, Proxmire tried and failed to get the Senate Appropriations Committee to delete the telescope's funding. Despite being chairman of the committee, Proxmire was outvoted 4 to 2. When the appropriations bill came up for debate before Congress, the issue of the space telescope was not even discussed. Instead, Boland tried on the House floor to delete funds for the Jupiter Orbital Probe, later dubbed Galileo, and was voted down.[69]

By the end of July 1977 NASA had chosen its contractors. The optical telescope assembly, holding the mirror, its frame, and the fine guidance sensors that would keep the telescope locked on target, would be built by Perkin-Elmer of Danbury, Connecticut. The telescope support systems module, including the gyros and power supply and body, would be built by Lockheed Missiles and Space Company in Sunnyvale, California. The two companies had bid $69 and $83 million respectively, considerably less than the $475 million that Congress had approved. Even when the cost was added to build the six science instruments that would sit at the back of the optical telescope assembly, on the surface at least it appeared in 1977 that the space telescope's budget had plenty of reserve cash, despite William Proxmire's and the GAO's concerns.[70]

4

Building It

In the spring of 1981, astronomer Kris Davidson took a two-month sabbatical from teaching at the University of Minnesota to travel to the Cerro Tololo Inter-American Observatory in the southern hemisphere. Located on top of the 7,200-foot-high Cerro Tololo peak in the high Andes mountains about 300 miles north of Santiago, Chile, this astronomical observatory was then one of the best places to go if an astronomer wanted to take a close look at some of the more unusual objects in the southern sky.

For Davidson, one of the more interesting of these unusual objects was the star Eta Carinae. "We already knew at that time that Eta Carinae was a really special object," he remembered. In the almost twenty years since Geoffrey Burbidge had first speculated that Eta Carinae might be the galaxy's next supernova, a handful of astronomers had managed to glean some additional basic facts about it. Apparently it was one of the galaxy's most massive stars, weighing somewhere around 100 times the Sun's mass, with surface temperatures greater than 50,000 degrees Fahrenheit. Further analysis of the star's spectrum indicated that its Homunculus was made of three separate shells of dust. One astronomer thought that the Homunculus was a ring of material tilted 70 degrees from our line of sight. Another noted that the growing Homunculus was likely caused by what he called "bipolar jets," jets of material shooting out from the star's poles.

Trying to figure out exactly how this all fit together with the images available, however, remained difficult. The best some astronomers had been able to do was to meticulously make repeated spectrograph cuts through the nebula, accumulating data on the motion and velocity of the gas to try to imagine what it really looked like. The resulting visualizations, as clever and thoughtful as they were, still remained vague, confusing, and unconvincing.

The biggest mystery for Kris Davidson, however, was the paucity of observations of the star. "It was just crazy," he said. "From 1920 to about 1980 or so hardly anyone ever looked at Eta, spectra, or images, or anything." He decided to use his sabbatical to go to Cerro Tololo and take a look.

Cerro Tololo in particular had one piece of new equipment that especially intrigued Davidson. Rather than record its images on film, the 1.5-meter telescope there had recently been equipped with a video detector called a Silicon Intensified Target (SIT) Vidicon. Developed by RCA, the SIT Vidicon had first been tested in October 1974 by a team of Caltech and JPL scientists led by James Westphal, using the 200-inch Hale Telescope at Palomar. The results had been startling. "We right quick took a one-second exposure, and here were stars fainter than we could see on the Schmidt [photographic] plate picture we had," Westphal later enthused. "We were whooping and hollering. I bet we could have been heard all over the top of the whole darn mountain. We were just mind-boggled, how sensitive the thing was."[1]

Since then, the SIT Vidicon had been improved. And because it was so sensitive, it could make its exposures quickly. "My idea was to go [to Cerro Tololo] and make better images then had ever been made before, just by making lots, bang, bang, bang, bang, bang, bang, bang," Davidson explained. With the vidicon he would take numerous images of Eta Carinae in a single night, afterwards sifting through these to find the best, which he could then enhance using computer processing.

Unfortunately, Davidson's effort was fraught with difficulties. On his one night of observation the sky was hazy. "We got what was claimed to be about one-and-half- to two-second seeing, which is lousy for an astronomical observatory." Worse, the only staff member at Cerro Tololo who understood how to operate the vidicon was out of the country. Davidson had to improvise, the results of which required weeks of processing to produce a usable picture.

Nonetheless, Davidson's 1981 image of Eta Carinae was at the time possibly the best ever taken of the star. It showed without question that

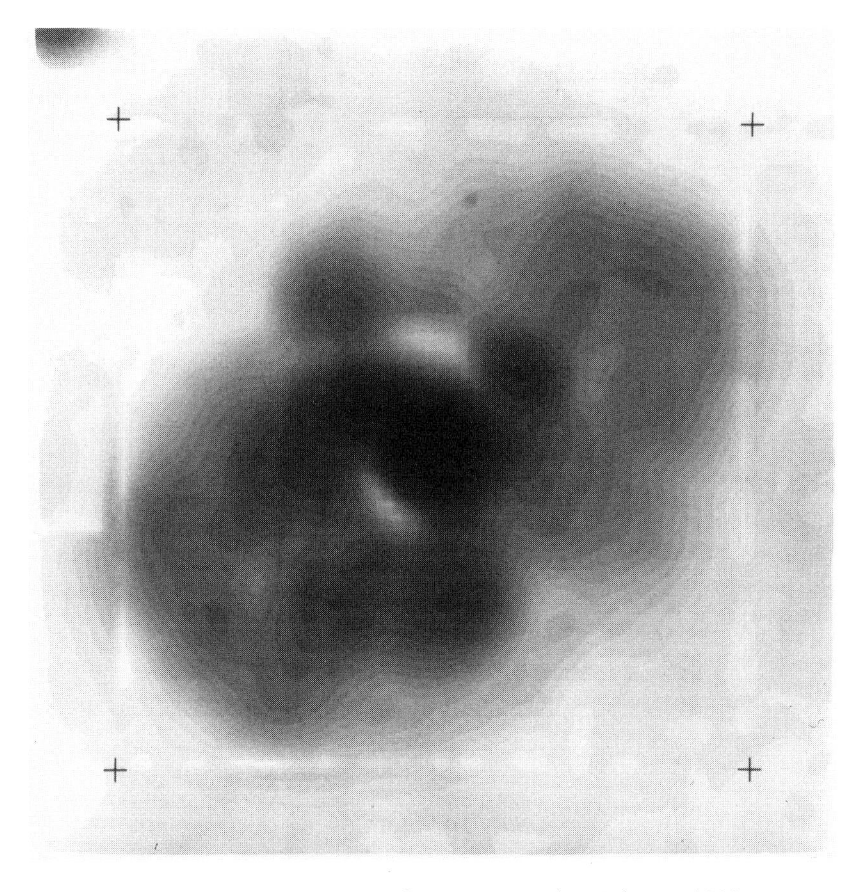

Fig. 4.1. Kris Davidson's SIT image of Eta Carinae, taken February 1981. (Photo by Kris Davidson, provided courtesy of Arnout van Genderen)

the Homunculus had grown since Gaviola had photographed it in 1945. More importantly, that growth had not been homogeneous, and suggested a variety of jets flowing in different directions.

Yet, despite these details, from Davidson's perspective the picture was simply not good enough to merit publication. Though he provided it to other scientists to use in their papers, he never published it himself. "I was disappointed with the quality. I didn't believe it was quite good enough." The image had artifacts in it that were not real, left over from the computer processing. And its overall fuzziness prevented him from learning anything striking or new about the star that hadn't been seen before. Though the picture suggested that the Homunculus was formed by jets, its resolution was too unclear to determine which blobs comprised which jets.[2]

■ ■ ■

His disappointment notwithstanding, Davidson's use of the SIT Vidicon detector was indicative of the revolutionary changes in technology that were both overtaking and being generated by the field of astronomy in the 1970s.

Not surprisingly, the space telescope had been at the fore of this revolution. Since the late 1960s, when the telescope's first design work was being done, it had always been assumed that the telescope would use some form of video technology to record its images and beam them back to Earth. Photography had been briefly considered, but rejected. Unexposed film had a very limited lifespan in space, quickly becoming fogged as it traveled through the radiation belts above the Earth. Moreover, no one could come up with a cheap and effective way to shuttle new and exposed film to and from space. Even in their most optimistic predictions about the frequency of shuttle flights, NASA officials could not imagine sending the shuttle up just to reload the film canisters inside the space telescope.

Television tubes seemed a hopeful alternative. The tubes could be used over and over, and the images did not need developing and could be transmitted from space relatively simply. Moreover, throughout the 1960s and early 1970s NASA had successfully used some form of video tubes in all its planetary and weather satellites.

Thus, in 1964 NASA's astronomy program under Nancy Roman began financing a series of studies out of Princeton University for building an orbiting optical telescope. At the time Spitzer had been pushing for the construction of what he called the Princeton Advanced Satellite to follow up on the Copernicus astronomy telescope he was building and expected to launch in 1971. With Roman's money, Spitzer put together a team at Princeton to develop a television camera using the Westinghouse Secondary Electron Conduction (SEC) Vidicon television tube that had been developed for the military.

Though the Princeton Advanced Satellite never flew, by the time the space telescope was under development in the early 1970s O'Dell's science working group had chosen Spitzer's SEC Vidicon as its principal detector. As Roman said years later, "We were locked into it at that time because there wasn't anything to compete with it."[3]

While the SEC Vidicon then seemed the best available choice for recording astronomical images in space, it had serious limitations. It did a very poor job of detecting light from the red end of the visible spectrum. It could only retain a charge for short periods of time, making long exposures difficult because the amount of light received would fail to accumulate. And the overall resolution across the field of view was inconsistent. On top of these issues, Westinghouse was having difficulty reliably manufacturing the tubes. As David Leckrone, the project scientist at Goddard, noted, "They went through runs of many tubes—up to 20, 23, or 30 tube starts—without having one that worked."[4]

During this same time period, the Jet Propulsion Laboratory (JPL) in Pasadena, California, was building several deep space planetary probes, including the Galileo probe to Jupiter, then called the Jupiter Orbiter Probe. Rather than use television tubes, the planetary scientists had chosen a different and very new technology for their cameras. Invented in 1970 by Bell Labs, charged coupled devices, or CCDs, were far lighter than tubes and had a good response in the red at wavelengths that are crucial for planetary research. Moreover, CCDs were excellent detectors, not only capable of accumulating light very precisely for long periods but able to pin down very accurately the relative location of each photon across the entire two-dimensional image.

Despite these advantages, CCDs had their own share of problems. The early SEC Vidicon had a large field of view, about 2 inches square, while the first CCDs were less than a tenth of that, producing extremely small images. CCDs also had to be kept very cold, −100 degrees Fahrenheit, in order to reduce the noise produced from the ambient heat, which would wipe out any faint astronomical target the CCD was trying to image. Worse, they had a very poor response in the ultraviolet part of the spectrum. For astronomers, this last defect was crushing, since the spectral signatures of the isotopes of many key elements, such as hydrogen, oxygen, carbon, nitrogen, silicon, and iron, could only be detected in the ultraviolet. Since the atmosphere blocked these spectral signatures, it was essential that the space telescope be able to study them once it got into orbit. The inability of CCDs to detect ultraviolet wavelengths therefore made them appear practically useless to the astronomers in the working group.[5]

The CCD, however, had one crucial advantage over the SEC Vidicon tube that had nothing to do with engineering: its commercial potential was unlimited. Many companies were willing to commit funds to develop the CCD, anticipating large profits using it in such popular items as digital and video cameras. The SEC Vidicon in turn was a dead-end technology, with the space telescope its only potential future customer.

From the time the CCD was first invented in 1970 through the mid-1970s, the technological developments accelerated. Step-by-step tricks were found to increase the size of CCDs, from arrays of 40 pixels to a side to 400 to a side. Ways were found for them to work at higher temperatures. The SEC Vidicon, meanwhile, showed little progress. Though NASA funneled millions in development money to Princeton, Spitzer's team just couldn't overcome the technology's limitations.

Nonetheless, despite these many misgivings throughout the early design stages, Spitzer's SEC Vidicon remained the choice for the telescope's main camera. "I think everybody really had a soft spot for Lyman Spitzer," noted Noel Hinners. "Lyman ought to be there as the PI." *[6]

At the October 1976 working group meeting at Caltech, however, the issue finally came to a head. This meeting was to be the last for the original science working group. Subsequently the project was to sort through the bids for the science instruments and from the winners reassemble a new science working group.

Once again—as had happened at almost every meeting for the last six years—the problems of the SEC Vidicon were discussed, this time during the afternoon session of the first day. Nancy Roman, who by this time was in favor of opening the camera to competition, gave a report on the situation, noting that in her opinion the development "had not progressed as fast as expected and there is now doubt that it can meet the minimum performance specifications." This was followed by several reports on the status of all the different types of detectors, including a de-

* PI stands for principal investigator, the scientist in charge of building and managing a specific scientific instrument.

scription of the CCD being developed by JPL for the Jupiter Orbiter Probe, Goddard's research into an intensified CCD for night vision work, and Westinghouse's SIT Vidicon work. When John Lowrance of Princeton presented his case for the SEC, he found himself under careful and intense questioning due to the continuing misgivings the scientists of the working group had for the vidicon's limitations.

After the first day's sessions ended there was a break before everyone gathered for dinner. Bahcall and Spitzer decided to take a walk, strolling through the parking lot just outside the new library.

They discussed the day's sessions. For Bahcall, the limitations of the SEC Vidicon had become unacceptable. Moreover, he was increasingly encouraged by the possibilities of CCDs. As a scientist he had to be objective in his decisions, and felt compelled to tell Spitzer that he could no longer support Princeton's bid to build the telescope's camera, unless it included CCDs.

For Spitzer, Bahcall's decision was crushing, directly threatening his status as the principal investigator on the telescope's main imaging camera. Not only would Bahcall's denial of support damage Spitzer's proposal badly, but his colleague's judgment that the Princeton commitment to the SEC Vidicon was flawed indicated that others might come to the same conclusion. "It affected him strongly. I remember it was an emotional discussion," Bahcall noted years later. "You know, I mean, emotional for the sort of discussion that you have with Lyman. He is not usually emotional."

Though Spitzer agreed to review the design of his camera and see if they could add a CCD to cover the red part of the spectrum, when the group reconvened the next day Bahcall nonetheless proposed that the telescope's camera be opened to competition. The group agreed, without opposition. NASA, under Nancy Roman's leadership, very quickly accepted the group's recommendation.[7]

Lyman Spitzer's team was suddenly out on a limb. They now had to compete against a new technology that many scientists on the project thought was clearly superior. Moreover, the competition was going to be fierce.

Among the many Caltech scientists who were attending this October meeting was James Westphal, the man who had led the development of the SIT Vidicon detector that Kris Davidson used to image Eta Carinae.

Though he didn't know it at the time, the vote to allow others to bid on the telescope's main camera was going to change his life.

Westphal, an impish man with a silver beard and enthusiastic manner, had no desire to work on the space telescope. He considered big government science projects as distracting and unpleasant. Instead, he preferred small hands-on experiments that he could do himself. Moreover, Westphal wasn't even an astronomer. In fact, he had come to the world of science from a most unlikely series of career choices, all inspired by his desire to avoid "doing some grungy things all my life."[8]

His childhood had been decidedly blue collar. His father had been an accountant during the Depression, but when Westphal was ten he quit that job and leased a gas station and parking lot in downtown Tulsa, Oklahoma, where the boy helped out by pumping gas and collecting parking fees during the summer.

Later his family moved to his grandparents' Arkansas mountaintop farm, where they raised cattle and produced dairy products. "It was all pretty much at a subsistence level," Westphal said in 1982. "Money was always a very serious problem." By the time he graduated high school there wasn't any money to send him to college. He left the farm and went back to Tulsa to get a job pumping gas.

At the same time he looked for any other work he could get, and eventually garnered a job working with a geophysical exploration crew at 35 cents an hour, working 60 hours a week, doing grunge work outside in the winter. "It was along in there that I decided I really wanted to go to college," he noted with humor in a 1982 interview. Working hard, he paid his way, earning a degree in geophysics in 1954 from the University of Tulsa.[9]

During and after college he worked in mineral exploration, searching for uranium and oil in remote jungles in Mexico and South America, working his way up the promotion ladder until he eventually became crew chief.

His big break came when he was offered the job to do geochemical research for the Sinclair Oil Company. The company's board had decided it needed a research department, and hired a whole crew of geologists, geophysicists, and geochemists. "The management . . . didn't have the slightest idea what should be done," Westphal explained. "They were

Fig. 4.2. Jim Westphal. (Photo courtesy of California Institute of Technology)

supposed to spend so many millions of dollars a year on research, so they said, 'Okay, here's a million dollars; go do research.'"[10]

Westphal's advantage, no matter where he went, was that not only did he have a natural mechanical genius for building things, but he also knew how to get people both above and below him in the chain of command to work with him. More importantly, he took a hard-driving but free spirit approach to his work. It had to be "fun" or "nifty" to attract him. For example, the primary focus of his research at Sinclair was to find unorthodox ways to find oil. To Westphal, that meant anything, from using aerial photography to gamma ray spectroscopy to gravity studies to the use of divining rods.

Westphal was also a superb salesman. One time he was scheduled to meet with the board of directors of Sinclair. Before the meeting he made a little mechanical toy to help break the ice. The toy was a little black box, about the size of a child's building block. On its side was a single switch. When you flipped the switch on, its top would open, and a small arm would reach out and down, and flip the switch back off. The arm would then quickly retract, leaving you where you started.

Before the meeting he showed this toy to some vice presidents, who forbade him from bringing it to the meeting with the big bosses, and threatened to have him fired on the spot if he did.

Westphal ignored the threat. "I took it in to the meeting in spite of what I had been told." He took out the toy and put it on the conference room table. Telling them that he was about to show them "a physical model of a manager," he flipped the switch. The board watched the arm come out and turn itself off.

There was then a long, deadly silence. Then, as Westphal remembered, "the person who broke was the comptroller who, you know, would normally be the stodgy guy in the system." Everyone in the room then started laughing.

Westphal's impish sense of humor won the board over. They ended up inviting him to join them for lunch in the executive dining room.[11]

During all this time, as much as he loved astronomy and space (he had built telescopes and played with rockets and explosives in high school and college), he had no interest in getting involved in NASA and the space science connected with the 1960s space race. Several times he was asked to participate in a planetary mission, and refused. "It's really not my style," he thought. He knew that if the planetary probe broke, he would not be able to get his hands on it to fix it. Besides, he just didn't like work where he himself wasn't tinkering with the equipment. With a typical NASA space mission, the best he could do would be to supervise the supervisors of other companies in the building of the instruments.

"For me . . . the object of the exercise," Westphal told historian David DeVorkin in 1982, "is just to have fun. I wouldn't [want to] miss the fun of putting together a circuit board or checking the hardware out, or trying to figure out why it doesn't work the way we thought it did. I find that fun."[12]

In the 1960s and 1970s, while Spitzer and O'Dell were getting the space telescope designed and funded, Westphal had been increasingly shifting his interests from engineering and geophysics to doing ground-based astronomy. In the process he improvised a whole new set of technological improvements to the field. In 1973 he made the first use of the SIT Vidicon for astronomy, employing the 200-inch Hale Telescope at Palomar to take some of the first electronic astronomical images ever.

Yet, though he was in the room when the space telescope's working group decided to open the competition for building the space telescope's main camera, he still wasn't interested in submitting his own proposal. Even when NASA released the RFP in 1976, announcing that it was looking for bids to build the telescope's camera, Westphal was not enthused. He had better things to do than to work on a big, bureaucratic space project. He was forty-seven years old, was comfortably situated at Caltech, and was free to divide his time between teaching and hands-on research, the things he loved.

About a month after the release of the RFP Westphal was sitting in his office at Caltech when he got a call. On the other end of the line was Caltech astronomer James Gunn, who asked Westphal if they could talk for five minutes. Westphal said sure. Less than a minute later Gunn walked into the room, closed the door, looked Westphal dead in the eye, and said, "We've got to build a wide field camera for the space telescope."

Westphal's reaction was typical. "You're out of your mind. . . . Why in the name of Christ would either of us ever want to get ourselves involved in that mess!"

"Because if we don't, we're not going to be doing astronomy 10 years from now."

"Come on, Jim, that's nonsense," Westphal answered.

"Tain't nonsense," Gun countered. "I'll show you."

Gunn proceeded to lay out his argument on the blackboard, showing how a space telescope was essential to the future of astronomy, how CCDs were the obvious tool for doing it, and how Westphal was the man who had already done much of the preliminary work making electronic detectors useful to astronomy. "[Gunn] had obviously been thinking about this, and he just had a list of 20 obvious things that you could do [in space] that you just couldn't do from the ground, no matter what," Westphal remembered. "The numbers were just mind boggling." As doubtful as Westphal was about working on a big NASA project, Gunn's arguments were compelling. "Okay, what do you want me to do?"

"I want you to be the PI."

Westphal answered, "You're crazy, Gunn!. You be the PI."

After throwing this hot potato back and forth several more times, they made a deal. They would list on the blackboard the six ideal people they wanted on the instrument team. If four agreed to do it, Westphal would be the PI. Otherwise, the job was Gunn's. They split up the names, and started calling people.

Since he was convinced no more than one or two would say yes, Westphal felt safe making the deal. To his horror, however, everyone he called was enthusiastic. For example, Westphal was sure astronomer Roger Lynds would say no. "Not only was he smart enough to know not to do it," he explained, "but it was even less his kind of thing then it was mine." Westphal called Lynds, got his wife BD, and left a message. In explaining to her the reason for his call, Westphal also expressed his confidence that Lynds wouldn't be interested. BD's answer was "Don't be too sure."

When Westphal's phone rang ten minutes later he was astonished to pick it up and hear Lynds's voice open the conversation with only one word. "He didn't say 'I'm Roger' or anything else," Westphal remembered. "He [just] said 'Yes.'"[13]

Unlike Spitzer's camera team, which had been committed to an older detector technology with a limited future and could only improvise clunky changes to what they already had, Westphal's team had the advantage of a fresh start, which meant they were open to any and all suggestions. For example, the solution to the CCD size problem was actually proposed by Bob O'Dell to Jim Westphal at the start of the October 1976 Caltech science working group meeting. O'Dell wanted the space telescope to have a detector able to produce images as large as about 2,000 pixels to a side. At the time, however, the largest CCD that Westphal's team at JPL had been able to create was around 400 × 400, far smaller than the SEC Vidicon detector.

O'Dell asked Westphal how development was going on the CCDs. One of his team members answered, "Maybe we could make an 800 by 800, but there it is. You guys want a 2,000 by 2,000, and it's just not going to happen."

O'Dell responded, "Why don't you split the image into four pieces?" In other words, O'Dell was suggesting that instead of trying to make a single large CCD, they make their detector out of a mosaic of four CCDs.

"Gee, that's an elegant idea," said Westphal. "I've never seen that before." Westphal immediately knew that by assembling an array of four 400 × 400 CCDs, each producing adjacent images that could be precisely stitched together later, a CCD detector could be built as large as the SEC Vidicon.[14]

Similarly, Westphal found a way to overcome the inability of CCDs to record ultraviolet light. He did a search of the literature, and discovered a very old spectroscopic technique whereby a photomultipler was coated with a material that fluoresced when hit by ultraviolet radiation, thereby allowing the photomultipler to detect it. Of the possible such materials, one named coronene seemed technically perfect for CCDs. Westphal did tests, even coating one half of a CCD with coronene and using it to image stars with the 200-inch Hale Telescope. At ultraviolet wavelengths only the coronene side could produce an image. "At that point we really knew we were home free," Westphal remembered.

To keep this discovery secret from other bidders, Fred Landauer of Westphal's team suggested giving coronene a code name, "mousemilk," for use in all documents. As Westphal remarked, "We needed a name to talk about without saying what it was, because we were really worried about somebody else thinking of the idea."[15]

When the time came in late September 1977 for O'Dell and the others to choose, the Westphal team instrument, dubbed by them the Wide Field/Planetary Camera in order to reassure planetary astronomers that it would be able to image the planets as well as distant astronomical objects, stood out far above the others. It was efficient, simple, and met the design specifications to the letter. The Spitzer proposal couldn't compete, and in fact, the decision wasn't even close. "What else could we do, in the light of the situation?" O'Dell noted years later. "The CCDs were coming on, came on even faster than I anticipated then. The SEC wasn't getting any better." In writing Bahcall shortly after, O'Dell noted, "The SEC-based proposal was panned so badly that I doubt that [the decision] can be protested." He then added, "This all very sad but is an inevitable result of the change in technology as the program dragged out, combined with Lyman's persistence in supporting John Lowrance and the SEC."[16]

For Spitzer, the loss was painful. His team at Princeton had spent years and a lot of NASA money researching the development of the SEC Vidicon. "I think we had no alternative but to submit a proposal based on the SEC, because of all the effort that had gone into it," Spitzer said in 1978. Moreover, at the time CCDs were still an untested and practically undeveloped technology. "We had the impression that NASA was being very conservative and would not be likely to approve something that was not in existence." Spitzer had hoped incorrectly that his older design would carry the day.[17]

Spitzer's public attitude toward this decision was stoic. By 1977 his retirement was pending. Moreover, he later rationalized how he "didn't have any enormous enthusiasm for the job because it didn't tie in with my own research interest particularly." During the 1950s and 1960s the bulk of his work had been spent trying to produce a controlled fusion reactor, while in the 1960s and 1970s he had been deeply focused on building and doing spectroscopic research with Copernicus, the last OAO satellite.

Nonetheless, the decision meant that Lyman Spitzer, after decades of work to get the space telescope built, had been bypassed as the principal investigator of the telescope's main camera. When the telescope was finally launched, he would not be one of the scientists at the heart of its operation and use.

Hope was not yet lost, however. Spitzer still knew that the telescope's operation center had to be based somewhere, and he had every expectation that that base would be at Princeton University.

Beginning in the 1960s, the operation of the space telescope once in orbit had been a hot-button issue among astronomers both inside and outside NASA. Goddard scientists and bureaucrats had repeatedly advocated that as the base for all NASA astronomy since the founding of the agency Goddard should operate the telescope. These Goddard scientists were quite alone in this position, however. In fact, among outside astronomers there was almost universal distrust and skepticism about Goddard's ability to do the job. Throughout the 1960s and 1970s scientists repeatedly expressed a wariness and dislike for the center and its employees. For example, when asked in 1966 for his thoughts on how Goddard treated outside scientists, Herb Friedman, one of the pioneer space scientists at the Naval Research Lab, was described as saying how "Those people at Goddard treat us like we are stupid. . . . They want to manage us. We at NRL will stay with the balloons and do other things rather than attempt to work with them."[18]

Bob O'Dell's own bad opinion of Goddard was based on the quality of the scientists the center had hired during the early boom years of the space race. In the early '60s many of Goddard's scientists had either not gotten their PhDs, or failed to finish their theses. In one case, for example, the Goddard center hired a former Wisconsin student who was not well thought of in Art Code's space lab at Wisconsin. Then, O'Dell's 1967 meetings with Leo Goldberg and Goddard's director John Clark helped reinforce his bad opinion of Goddard's scientists. As a result, almost as soon as he became project scientist, O'Dell campaigned for the establishment of an independent "Space Telescope Science Institute," as he dubbed it, to operate the telescope outside of Goddard. To put it simply, O'Dell did not want the space telescope controlled by the civil servant astronomers working at Goddard.[19]

This schism between Goddard scientists and the rest of the astronomical community was most starkly illustrated at an October 1974 meeting of the science working group. At this meeting, held in Cambridge, Massachusetts, a heated debate took place about who should control the telescope's operations after launch. To illustrate clearly who stood where on the issue, O'Dell decided to poll everyone. Only the Goddard and

NASA employees were willing to consider letting Goddard run the telescope. They argued that giving control of the telescope to an independent "institute" created the uncomfortable situation where NASA, the agency responsible for managing the project, had no control over its management.

Everyone else, however, either opposed Goddard or was doubtful. As O'Dell noted in the meeting's minutes, "The consensus of the working group appears to be an autonomous, non-NASA institute."[20]

Despite the broad skepticism expressed by the astronomers at this 1974 meeting, it took several more years to convince NASA's management that the telescope's operation should be placed under the control of an independent institute. It wasn't until the release of a National Academy of Sciences study in 1976, chaired by Donald Hornig, former science advisor to President Lyndon Johnson, that the idea of a Space Telescope Science Institute was finally considered acceptable by NASA management.[21]

Even Spitzer, who had rarely expressed any public criticism of Goddard, noted several times during the extended debate how the astronomical community was skeptical of NASA.[22] Nonetheless, though he supported the idea of this institute, the issue to him was initially irrelevant. His group at Princeton was building the SEC Vidicon that was to be the telescope's camera. Whether or not the telescope was managed by Goddard or an independent institute, he expected to be the camera's principal investigator and thus operate the telescope's camera, with a guest observer program run out of Princeton under his supervision (much as he had done for Copernicus).

When he lost the camera to the Westphal team, however, the issue of getting the institute established at Princeton suddenly became paramount to Spitzer.

Getting the institute job was not going to be simple, however. Specific universities like Princeton didn't really have the option of bidding on their own. Following World War II it had become traditional and expected that one of a handful of university consortiums, usually made up of anywhere from a dozen to almost a hundred universities, would operate facilities such as the Space Telescope Science Institute. Of these consortiums, Associated Universities, Inc. (AUI) was the first, formed in 1947 to manage Brookhaven National Laboratories on Long Island, New

York. Later, in 1956, AUI took over operation of the National Radio Astronomy Observatory, first established in Greenbank, West Virginia, and later expanded to include radio telescopes in New Mexico and elsewhere. Later consortiums included the Association of Universities for Research in Astronomy (AURA), which began in 1957 managing the Kitt Peak National Observatory in Arizona; the Universities Research Association (URA), which since 1967 has operated Fermilab in Batavia, Illinois; and the Universities Space Research Association (USRA), formed in 1969 to manage the Lunar Receiving Laboratory, which would store and dissect the lunar rocks that the Apollo astronauts brought back from the Moon.

A single university like Princeton could not compete against such consortiums, which had both the right management experience and the collective support of the many universities they represented. Therefore, rather than submit his own independent bid, Spitzer and John Bahcall, who supported Spitzer's effort to bring the telescope's operations center to Princeton, worked out a strategy to have Princeton named as the location for the Space Telescope Science Institute on as many different consortium bids as possible. If every bidder named Princeton, then Princeton would be chosen no matter which consortium won the bidding.

During the months leading up to the bidding, five organizations showed serious interest: AUI, AURA, USRA, URA, and one private company, Battelle, which had made it its business doing this kind of work, for example running the Hanford site of the Pacific Northwest National Laboratory for the Atomic Energy Commission since 1965. Together, Spitzer and Bahcall pitched Princeton to all five.

The AURA executive committee, led by John Teem but also including Art Code and Margaret Burbidge, weighed its options very carefully and decided against picking Princeton. At the time AURA's reputation was mixed because of management problems it had had in running the Kitt Peak National Observatory in Arizona. To overcome this negative impression the committee believed they needed to make their bid as unique and outstanding as possible. Thus, choosing Princeton would leave their proposal indistinguishable from all the others.

Instead they considered finding a site close to Goddard. When they approached the University of Maryland at College Park, they were told, no, the university was already being courted by USRA. It was then that

Teem and Art Code, who was going to be the institute's acting director under their bid, talked to astrophysicist Arthur Davidsen at Johns Hopkins University in Baltimore, Maryland. Davidsen's first reaction was mixed. "It would be great fun and wouldn't it be a wonderful thing, but we are way too small to have any chance." Very quickly, however, Davidsen changed his mind, especially when others in the Johns Hopkins astrophysics department became enthusiastic over the idea. As a result, Johns Hopkins became AURA's choice, with Art Davidsen taking a leading role in writing its proposal to NASA.[23]

The scientists at URA, meanwhile, had liked Spitzer's and Bahcall's pitch and had recommended that Princeton be named in their proposal. However, their managers overruled this recommendation, insisting that they wanted the institute located close to their main base of operations at Fermilab in Illinois.[24]

The remaining three bidders, AUI, USRA, and Battelle all picked Princeton. In fact, the Princeton proposal was so good that it convinced USRA to switch to Princeton from their original team-up with College Park. AUI in turn favored the Princeton location so much they decided to choose Bahcall for their first institute director, though they did not include this fact in their proposal.[25]

Thus, when the five proposals were reviewed by NASA in September 1980, the odds looked very good for Spitzer. Not only did three of the five proposals list Princeton as the preferred location, there was Spitzer's decades-long relationship with NASA and his long and successful track record running Copernicus and the Plasma Physics Laboratory at Princeton. "I thought Princeton had really a very good chance of being the site of the Space Telescope Science Institute, with one consortium or another," Spitzer said in 1984.[26]

When NASA reduced the bidders to two finalists, AUI and AURA, however, Spitzer's chances shrank. Worse, of the two, AURA's proposal was far more impressive to NASA managers. Unlike AUI, AURA had included an outside contractor, Computer Science Corporation, for operations and software management at the institute. This fit NASA's wish that the institute be as flexible as possible, able to change its staffing quickly as its needs changed. AURA also proposed a much smaller institute than AUI, which matched very closely the wishes of the managers at NASA headquarters.

In a sense, the dark horse nature of AURA's bid, with Johns Hopkins University as its location, became a key factor in winning the competition. Driven by a sense of insecurity, AURA's proposal was longer, more detailed, and more thoroughly researched than its competitors.

AUI, meanwhile, had taken too much for granted. "AUI could have done the best job," O'Dell remembered. "That is, they knew they could do the job, and [so] they didn't work hard in preparing their proposal."[27] When NASA's evaluation board compared AUI's proposal with AURA's, the choice was a no-brainer.

Art Davidsen was at a meeting at Goddard soon after the decision had been made. "The rumors had been out that the announcement was going to be any hour now," Davidsen remembered in 1984. "There were people there from NASA that knew that were just smiling at me, you know, now you're going to be happy soon."

He called back to Johns Hopkins to find out if they had gotten any news, and discovered that Hopkins was the winner. Someone also gave him the message that John Bahcall had called. The two men were friends, and Davidsen figured Bahcall had called to congratulate him. He in turn felt compelled to call Bahcall to express his sympathy.

Bahcall had not heard the news, however. "I had the unfortunate duty of being the first one to tell him," remembered Davidsen. "I could hear John . . . just sort of appearing to . . . collapse from the phone."

Bahcall later claimed otherwise. "I was the most relieved person in Princeton." No longer was he faced with the possibility of giving up research to become a manager like O'Dell. He was once again free to do science full time.[28]

Spitzer, meanwhile, once again found himself outside looking in. After more than four decades of effort, the man who had first proposed that the United States launch a large optical telescope into space would have no official part in its use. "I was disappointed," he remembered in 1984. "I sometimes ask myself why I worked so hard."[29]

Ironically, the astronomer who ended up running the institute for the next ten or so years was someone who had had little to do with the space telescope—in fact, he had originally opposed its construction. Riccardo Giacconi, the man who built the first x-ray space telescope and discovered the first x-ray emitting objects in space after Nancy Roman had rejected his proposal in 1960—and for which he won the Nobel Prize in 2002—had lobbied against funding the space telescope in the mid-1970s when O'Dell, Bahcall, and Spitzer were pushing Congress for funds. Instead, Giacconi had wanted Congress to build more orbiting x-ray telescopes.

When the space telescope was approved, Giacconi found himself without funds to build more x-ray telescopes. "I was not being fully used," Giacconi explained. His research career was stymied. By taking over as head of the telescope's operation he could at least participate in one of the most important space astronomy projects of his time, and bring to it his passion for quality and excellence.

Thus, in June 1981 he was appointed the first permanent director of the Space Telescope Science Institute, relieving Art Code from his acting position.[30]

■ ■ ■

Even as Lyman Spitzer was slowly finding himself eased out of the space telescope project, the construction of the space telescope moved forward fitfully, fraught with its own moments of panic and heartache. Leading the way in these troubles was the company Perkin-Elmer, located in Connecticut and given the job of building the telescope's mirrors, the frame to hold them, and the fine guidance sensors that would be used to point the telescope more precisely than had ever been attempted.

The technical and scheduling difficulties began almost as soon as the blank mirror glass was shipped in December 1978 from Corning in New York to the Perkin-Elmer plant in Wilton, Connecticut. First there was the "teacup" affair. On March 27, 1979, having completed the first preliminary surface grinding of the 94-inch primary mirror, Perkin-Elmer technicians removed it from the grinder to inspect it. To their horror, they discovered a network of tiny cracks shaped like a teacup about a quarter inch in diameter and visible on the inside surface of the mirror's front plate and near its outside perimeter.

For the next two weeks Perkin-Elmer's engineers argued about what to do. Their fear was that if they did nothing the cracks would spread, ruining the mirror. To prevent this the usual practice in the case of a small network of cracks like the teacup was to drill it all out, leaving behind a quarter inch plugged hole from which no cracks could spread.

Since the hole was small and near the edge of the mirror, it would have little effect on the mirror's light-gathering capability.

The mirror blank itself was actually two 2-inch faces fused to an intermediate grid. The grid gave the faces rigidity while saving weight. Drilling the teacup out from the top face carried its own risks. The drill process could put pressure on the cracks and force them to spread before they could be cored out. To drill from below where the cracks were located was risky also, since it would scatter glass dust throughout the mirror's interior grid, dust that once in zero gravity could drift about and settle on the mirror face, ruining its vision.

Finally, after weeks of discussion the engineers agreed to drill from the top. To everyone's relief the operation worked, removing the teacup without producing further damage.

The delay caused by removing the teacup only compounded the scheduling and management confusion at Perkin-Elmer. During the weeks-long crisis, Bill Keathley, the telescope's project manager at Marshall during these first years of construction, made several trips to Perkin-Elmer to monitor the situation. He quickly noted how the only way he could get managers there to do anything was with "a great deal of government pressure and direct involvement." Worse, Keathley worried, was that everything was done in a "panic mode." Similarly, when O'Dell went to Connecticut later that year in November to review their progress, company executives were unprepared to discuss anything but the primary mirror with him. What O'Dell saw was "confusion."[31]

Much of Perkin-Elmer's problems had to do with money. They, like the NASA bureaucrats and their compatriots at Lockheed, had seriously underbid the cost of building their portion of the telescope. As one company official reportedly said, "[Perkin-Elmer] had to lie to get the contract. NASA had to lie to get the money." Thus, neither had sufficient resources to do what needed to be done.[32]

Magnifying these woes was the continuing friction between the managers at Goddard and Marshall. The RFP for the Space Telescope Science Institute was being written, and many at Goddard were resisting O'Dell's effort to keep telescope operations away from Goddard. In addition, both O'Dell and Keathley were unhappy with how Goddard was managing Westphal's team at JPL.[33]

Considering this situation, it therefore wasn't surprising when on January 21, 1980, Bill Keathley announced that he was leaving the space telescope project to take a job at Goddard. "He got out at the right time," noted O'Dell years later.[34] The space telescope would have to find itself a new project manager.

The man whom NASA officials favored for the job was Fred Speer, who had been running the High Energy Astrophysical Observatory Project for Marshall for the past eight years. What made Speer appealing to them was how he had taken that project, over budget and out of control, and reshaped it so that it could get built for much less money. Speer, originally from Germany, was unlike the other German engineers who worked at Marshall in that he had not been part of the V2 project during World War II and had not come over with Wernher von Braun right after the war. Instead, he had been an assistant professor of physics at the Technical University in Berlin during the war and afterward. In this position he had known von Braun casually, but had not worked with him.

Space and rocketry fascinated him, however, and in 1955 he wrote a letter to von Braun, asking if he could come to the United States and join his team. To Speer's delight von Braun offered him a job. Speer began by evaluating the flight tests of the Redstone rocket, which eventually sent Alan Shepard into space. Then he moved on to the Saturn family of rockets, developing operations support centers in Marshall and Kennedy to gather and analyze the flight data from each launch.[35]

In 1971 he was appointed project manager of the High Energy Astrophysics Observatory program, called HEAO. At the time the HEAO program included two very large and sophisticated 20-ton x-ray satellites. Six months after Speer took over, however, Fletcher pulled HEAO from the budget, saying it was too expensive. Instead, he gave Speer and the scientists one year to regroup, and after much tearing of hair they came up with a much cheaper program, involving three satellites whose total weight equaled one of the original two HEAO probes. Over the next eight years Speer supervised this program, keeping it under budget and on schedule. His success made him a 1977 winner of *Aviation Week and Space Technology*'s Laureates Award "for turning the high-energy astronomy observatory (HEAO 1) satellite program from a canceled mission in 1973 to a major scientific success."[36]

Despite this success, the scientists working on the space telescope were very leery of Speer as the project manager. His hard-nosed approach in getting HEAO built frightened them, especially since they knew the cash-poor state of the space telescope's budget. "In the HEAO program," O'Dell said, "all who had worked with him there at Marshall were against him, saying that he was a person utterly unsympathetic to the science parts and would be bad news for the space telescope project." When it came time for O'Dell to make his recommendations, he admitted that Speer could do the job, but also recommended someone else as his first choice.

Nonetheless, Bill Lucas, the director of Marshall, picked Speer. As Noel Hinners told John Bahcall, "Headquarters perceives no one else at Marshall as qualified."[37]

Almost immediately Speer began an end-to-end review of the project, and quickly realized how over budget and behind schedule it was. As he told O'Dell just weeks after taking over, "We have already blown our launch date."[38]

O'Dell quickly discovered that Speer's methods were decidedly different from Keathley's. While Keathley had what O'Dell called a "very loosey goosey" style of management—"promise you this and promise you that"—Speer was clear-cut and to the point.[39] If he saw a problem he attacked it, sometimes proposing radical solutions that no one else was willing to consider. For O'Dell this directness was at first refreshing, considering the tightness of the telescope's budget. As time went on, however, Speer's brutally direct approach frustrated him, since it seemed to O'Dell that Speer was putting the budget ahead of building a useful telescope. For example, having recognized that the project faced serious cost overruns, Speer began looking for things that could be cut. First he asked O'Dell if the large hinged door on the front of the telescope could be deleted. Then he suggested eliminating the small actuators attached to the back of the mirror to permit coarse focus adjustments. Next he proposed deleting a number of instrument tests that were planned at Goddard. Finally, he suggested shortening the list of units being designed for easy removal and replacement by astronauts while in orbit.

O'Dell saw most of these cuts as penny wise and pound foolish, and was somewhat effective in fighting them. After much debate he convinced Speer that the aperture door was essential, since its removal would not save much money, while its lack on the orbiting telescope would reduce the amount of possible science. Without the door to act as an additional sunshade, the telescope would only be able to look at about 50 percent of the sky, instead of 82 percent. As for the actuators, though O'Dell agreed that they were probably less crucial, he argued successfully that the cost of removing them at this point in construction was about the same as including them.[40]

O'Dell was less successful in saving the removable units and the tests of the scientific instruments at Goddard. Though the telescope's original design had called for the majority of the telescope's instruments to be maintainable in orbit, this kind of flexibility was expensive to build. In his effort to save the budget, Speer got the number of replaceable units reduced. Similarly, he eliminated many of the tests of the science instruments at Goddard, arguing that these same instruments would be tested again when they were installed in the telescope at Lockheed in California. From his perspective, it made no sense to do these tests twice.

Even these cuts weren't enough. Throughout the spring of 1980 Speer put great pressure on O'Dell to find additional cost savings out of his science instrument budget. Nor was Speer alone in applying pressure to O'Dell to trim the budget. Frank Martin, who at the start of 1980 had replaced Bland Norris as O'Dell's immediate boss at NASA headquarters, was demanding that Speer and O'Dell get the project's budget under control, no matter what it took.[41]

Meanwhile, O'Dell had his own personal frustrations. His marriage of more than twenty years had ended in divorce in 1979, the previous year. Then, when Ernst Stuhlinger had retired in 1976 O'Dell had replaced him as associate director for science at Marshall, adding this second job on top of his work as the space telescope's project scientist. "Project scientist was not a full-time job," O'Dell explained. "It just didn't take all of my time." The new job had two functions: steer new science work to the center, and review and guide the science work already there. However, though the job gave O'Dell some responsibility, it did so without giving him any authority. While Stuhlinger had had excellent ties to the many other German engineers at Marshall, being German himself, O'Dell did not have this advantage. "I was coming in as an outsider."[42]

Amidst these pressures, O'Dell began his own review of the six space telescope science instruments, and began making demands on each. In particular, he focused on the High Resolution Spectrograph (HRS) being built at Goddard. The principal investigator of this instrument was Jack Brandt, who had taken over the Solar Physics branch at Goddard in 1967 (after O'Dell had recommended him for the job), and then became head of Goddard's Laboratory for Astronomy and Solar Physics in 1977, leading among other things the effort to build the Solar Maximum Mission.[43]

Relations between Brandt (a Goddard man all the way) and O'Dell (who did not have much faith in Goddard scientists) had never been smooth. Brandt, whom Ed Weiler, Nancy Roman's successor as NASA chief astronomer, once referred to as a "nattering nabob of negativism," had for example openly fought O'Dell's effort to keep telescope operations out of Goddard. He also predicted (rightly it turned out) in a meeting in early 1980 that the telescope's then-scheduled launch date of 1983 was absurd, and that the telescope would likely not launch before 1988.[44]

Now their working relationship became downright hostile. On April 23, 1980, at a cost review meeting with Speer, O'Dell noted that Brandt's spectrograph was behind schedule, caused partly by what O'Dell considered a lack of time commitment to the spectrograph project by Brandt.[45]

As a result of this and later discussions, when the science working group met on May 5 and 6 at Goddard, both Speer and O'Dell pressed Brandt hard over the instrument's budget and scheduling problems. In particular they were worried about what Brandt called "hybrid electronics packages." Though Brandt's team had managed to get these units built, they had had trouble sealing them in accordance with specifications. Brandt, however, was unconcerned, claiming the issue was "only a question of semantics."[46]

The day after the meeting Speer called Brandt and told him that it was his intention to delay the launch of any science instrument should it develop serious problems. And from what Speer had seen the day before, the spectrograph was having serious problems.[47]

O'Dell meanwhile was reviewing the spectrograph design to see if it could be simplified and thus be finished for less. One of the spectrograph team's other difficulties was the manufacture of what was called an echelle grating, used to split the light like a prism. After reviewing the situation O'Dell recommended a simpler design, noting that "the high resolution mode ($R = 10^5$) . . . is clearly a secondary function of this instrument" and could be replaced by something simpler "for essentially no cost." By replacing this mode, O'Dell explained, the problem grating could also be substituted by an available "replica grating," thereby eliminating the grating problem.[48]

O'Dell, however, had gotten his modes confused. In a blistering response, David Leckrone, Goddard's project scientist for the space telescope, explained that it was the $R = 10^3$ mode that was expendable, having been added later "on the condition that it would be of negligible cost and impact." The high-resolution $R = 10^5$ mode, however, was "an important part of the original HRS proposal to NASA and which has been a major design driver from the beginning." Leckrone also noted in great detail how the problems with the echelle grating were hardly insurmountable, and in fact, were being dealt with successfully.[49] O'Dell's long-term misgivings about the scientists at Goddard had allowed him to badly misinterpret the situation.

Even as these battles were going on, O'Dell had been considering other more severe options for meeting Speer's and Frank Martin's cost-cutting demands. He pondered how ground-based telescopes operated. Astronomers would build the telescope first but delay the construction of some instruments, like spectrographs, until there was enough money to pay for them. As long as the telescope could do optical observations full time, its operation would not be impaired. "Since we were gong to have access to the shuttle and a long life for the observatory," O'Dell thought, "it made perfectly good sense to me to do it the same way here."[50]

The idea of delaying the space telescope's two spectrographs appealed to O'Dell from another perspective: the HRS wasn't the only spectrograph having budget and management problems. The Faint Object Spectrograph's (FOS) principal investigator, Richard Harms, was a young man who apparently was initially unprepared to do the job. In an early project review in July 1978, one engineer noted with amazement how Harms was "absolutely nonplussed when during a discussion [he] admitted not opening his mail. . . . This is crazy, and someone should read him the riot act." By the end of 1978, FOS's budget and scheduling problems had gotten worse. George Levin, the project manager at Goddard, had made a tour of all the science instrument operations, and was astonished to find Harms on vacation in Australia when he arrived at the University of California to review the situation and demand changes. A university review team agreed that major management changes were needed, and after much discussion appointed another team member, Bruce Margon, as deputy principal investigator to assist Harms through this rough period.[51]

Throughout the spring O'Dell discussed the idea of delaying some instruments with both Speer and other NASA bureaucrats, without being too specific. When at a space telescope staff meeting on July 14, however, Speer described the project's "cost picture [as being] even worse than thought," O'Dell decided he had no choice. On July 17, he pulled Speer aside and offered him a plan whereby the telescope would be launched initially with just the cameras and photometers. The full complement of spectrographs would be added later. Speer liked the idea, and immediately began to incorporate this plan into his presentation for the next budget review meeting, scheduled in about a week.

Unlike in 1976 when the scientists proposed shrinking the mirror from 120 inches to 94 inches, however, O'Dell had not polled the other scientists in the science working group about his idea of delaying the two spectrographs. He had not even told the scientists building the two instruments, leaving them in the dark about the possibility that their projects might get delayed by years. It was a tactical and strategic error of the first order, leaving both him and Speer vulnerable to attack.

The reasons for this error were complex. Not only had his relations with Brandt deteriorated, but O'Dell no longer had many allies in the science working group. Most of the people with whom he had worked and become close friends during the early funding and design battles were either not part of the instrument teams or, if so, were very minor players. Lyman Spitzer was gone. George Field was gone. Nancy Roman was gone, having just retired. Margaret Burbidge was on the Faint Object Spectrograph team, but was consumed with her own budget and scheduling troubles. Of those chosen for the working group, many were very inexperienced when it came to operating research satellites in low earth orbit. As O'Dell wrote to Bahcall right after the instrument contracts were awarded, "Only you and Bob Bless return," adding, "We need your good advice."[52]

As one of the few individuals there from the beginning, O'Dell began taking a more authoritarian approach toward running the science working group meetings. O'Dell's dominance amazed Charles Pellerin when he attended his first science working group meeting after becoming deputy to Frank Martin. "I never saw a person with strength of personality take such a powerful group of people and just absolutely control their conversation; but that's what he did." David Leckrone was less kind. "If anyone in the group didn't agree with what he wanted them to agree to or argued with him or tried to turn things in a different direction or tried to assert any kind of independence, he would turn red in the face, he would slam his fist on the table, he'd call a coffee break and stalk off."[53]

O'Dell himself had hints of this situation, such as the time, the same month he was fighting Brandt over the spectrograph, that he met a co-worker in the hallway and noted how she was "*very* unfriendly [to me]. When I remarked on her preoccupation, she said it was about weighty issues that *I* knew *all* about."[54]

Unfortunately for O'Dell, his tough management approach was doing little to fix the project's problems, while putting a wall between himself and everyone else, a situation that could not work in his favor when he proposed delaying the spectrographs at a cost review at Marshall on Saturday, July 26, 1980. To both Speer's and O'Dell's discomfort, the suggestion was met with unmitigated horror and disapproval, even among the same NASA bureaucrats who had been demanding that they find a way to reduce the project's cost overruns.

Over the next two months O'Dell was faced with a barrage of angry criticism from all quarters. At the September science working group meeting at Marshall Brandt attacked him, accusing him of improper conduct. Then the rest of the working group piled on, complaining about their lack of inclusion in the decision.[55] A week later Frank Martin, who had demanded that Speer and O'Dell cut costs, called O'Dell into his office and ripped into him, telling him he had betrayed the scientific community and that his plan was indefensible. "He really chewed my ass," O'Dell remembered in 1985.[56]

Even some of O'Dell's past allies were furious. Margaret Burbidge, incensed at the threat to her Faint Object Spectrograph, called O'Dell and told him it seemed unreasonable to take most of the cuts from the science teams when Lockheed and Perkin-Elmer were having most of the budget problems.[57]

Only Bahcall offered O'Dell any support, writing him a short note saying "I'm glad you're in charge! The approach and decisions you have taken seem very reasonable and even wise."[58]

As O'Dell noted in a letter to Bob Bless, "I felt a bit 'abandoned to the wolves.'"[59]

For O'Dell, that Saturday cost review meeting—later labeled "Black Saturday" by people in the project—was the final crushing blow to his influence in the project. Not only were his opinions never again treated with the same respect by the science working group, but he himself lost heart, becoming less dedicated to the project. Though still helpful, after Black Saturday O'Dell was no longer willing to go out on a limb to fix any new problem.

He began to look for work elsewhere. Finding work, however, turned out to be more difficult than he expected. He had made so many enemies in NASA that when the time came to move on there was nowhere in NASA for him to go.

Too many people at Goddard disliked him. He called George Pieper, asking him if the job offer from five years earlier was still open. After polling his coworkers Pieper told O'Dell that he was no longer welcome at Goddard. His passionate efforts to keep the scientific control of the telescope out of Goddard's hands had "rubbed too many people the wrong way."[60]

Nor could O'Dell stay at Marshall. It was an engineering center, and there was little for him to do there as an astronomer. Further, once the telescope was launched the center would have no say on its use, with Goddard taking over the telescope's day-to-day technical operation.

He couldn't move to the Space Telescope Science Institute in Maryland. As a government employee who had helped create the institute, he was forbidden by law from ever being employed by it.

As a scientist, his access to research facilities was also shaky. Unlike the other members of the science working group who were in charge of and building specific instruments, he had no guaranteed time on the space telescope. When it was launched, he would have to get behind every other scientist in the world as they lined up to use it. And since he was no longer affiliated with any major university or observatory, his access to other ground-based telescopes was going to be as limited.

When he started to look for work in the academic world, however, he discovered that his stock as a scientist had also fallen. Astronomer Don Osterbrock, who had been one of O'Dell's advisors when he was a graduate student at the University of Wisconsin, explained to him that getting a teaching position would not be easy, that most of the present generation of astronomers saw O'Dell not as a scientist but as a manager. After several months of searching, the best he was able to find was a position at Rice University in Texas, in what, as O'Dell described in 1985, "is not a major department. It's not like the scholarly institutions I was at before."

From what O'Dell could see, the decade he had devoted to getting the Hubble Space Telescope built had destroyed his career. Even the collapse of his marriage could be attributed, at least partly, to the project. "I don't know if I would have divorced if it hadn't been all the business of living in a small town in Alabama, the problems I had with my kids which I related to that, the stress that that produced between my wife and I."

It was a bitter pill to swallow. "I somewhat look at myself as a casualty of the space telescope," he said in 1985. "I lost a lot personally."[61]

On September 7, 1982, ten years and one week after he started as the project scientist at Marshall, Bob O'Dell moved to Houston to take up work as a professor of astronomy at Rice University. Though he initially kept the job as the space telescope's project scientist, doing it long distance and part time was unrealistic. On October 1, 1983, he was finally replaced by astronomer Bob Brown.

■ ■ ■

Even as these battles were unfolding in Huntsville, Beltsville, and Washington, DC, other events were occurring in Connecticut—unbeknownst to practically everyone—which were to have far more significant consequences for the future of the Hubble Space Telescope.

Only weeks before Black Saturday the rough grinding of the telescope's eight-foot-wide primary mirror had finally been completed, and in mid-June Perkin-Elmer had shipped it from the company's Wilton plant in Connecticut to its more sophisticated operations in Danbury, Connecticut. There, using very precise computers, the company planned to fine polish the mirror, giving it a surface so precise that, once installed in the telescope, it would make it possible for a person to distinguish the difference between a dime and a silver dollar from a distance of 40 miles.[62]

Bud Rigby, in charge of polishing the mirror, was under incredible pressure, far more than even this kind of high-pressure project would have normally produced. The budget restraints and the difficulty of the job had put the entire mirror-polishing seriously behind schedule. Rigby's team had expected to complete the rough grinding of the mirror, which was originally shipped to Wilton in late 1978, in only nine months. Instead, it had taken them almost twice as long.

Besides the six weeks they had lost due to the teacup affair, there were other problems with the glass that Corning had delivered. In joining the blank's top and bottom faces to the interior gridwork, the outside edge of one face had improperly fused to the gridwork in one spot. Because such an imperfection was a potential source of uneven stress that could damage the mirror during launch, Corning engineers had tried to cut out the fused sections, leaving the mirror with several straight-edged grooves that, to Rigby, were an even worse problem. Straight grooves could easily become the source of new cracks that could quickly spread. Before he could begin the rough grinding, Rigby's crew spent weeks using acid to burn off the sharp edges. They then cut out the improperly fused sections with dental tools.

By the time the primary mirror arrived in Danbury for fine polishing, Rigby and his team were under considerable pressure to get on with the job. In fact, the pressures were so intense that Rigby often found himself working until the wee hours of the morning, and rather than drive home, he eventually decided to have a cot installed in a trailer in the parking lot so he could sleep there.

Before his team could begin fine polishing, however, major changes had to be made to the equipment they would use to precisely measure the mirror's shape as they polished it. For the last few months Rigby's team had labored hard making a five-foot test mirror for NASA, proving that their polishing equipment could produced a mirror with a surface smooth enough to meet the space telescope's specifications. To make sure they were polishing this mirror correctly, they measured its shape using something called a reflective null corrector. Made up of two mirrors and one lens, this null corrector was far more precise than earlier designs, able to measure deformities on the face of the mirror as small as ten billionths of an inch. To do this the light from a laser was shone through the null corrector onto the primary mirror and then reflected back into the null corrector. Normally, the wavelengths from the two beams would cancel each other out and nothing but black would be seen (hence, the use of the word "null" in the device's name). Instead, the null corrector's lens was carefully adjusted so that the wavelengths were slightly offset and produced alternating strips of bright and dark, what optical engineers officially call an interferogram but often dub simply as "fringes." These fringes were like the contour lines on a topographic map: if the primary mirror was curved properly, they would look like a series of straight horizontal lines. If the curvature was imperfect, or incorrect, the fringes would twist and bend, just like the lines on a topographic map.

In charge of the null corrector's setup and operation was Lucien Montagnino. Careful and meticulous, Montagnino was also known to be possessive and controlling. He didn't like others questioning his work. As David Burch, one of the company's quality control officers, noted, "He absolutely was an extremely dominant force."[63]

Montagnino's demanding personality—combined with his extensive knowledge of optics—had served him well, however. He had been in charge of the null corrector during much of Perkin-Elmer's successful defense contract work, a fact that worked to the company's advantage when it bid on the job to build the telescope assembly. While Kodak's solution was to build two mirrors, testing both with different equipment and then picking the best, Perkin-Elmer had a design that had already done the job more accurately, more efficiently, and more quickly. In fact, it was Perkin-Elmer's new design of this old technology that had been one of the main reason's the company had gotten the job.

So, in July 1980, even as the astronomers and NASA managers were fighting over the bones of their money-starved telescope, Montagnino's testing crew hurried to get the null corrector reconfigured from measuring the shape of the five-foot mirror to measuring the shape of the eight-foot primary mirror. To do so, they had to reposition the corrector's two mirrors, then readjust the position of the lens so that it was exactly 58 centimeters (22.83 inches) below the focus point of the lower mirror.

Moreover, the null corrector was located in a somewhat inaccessible place, in a small room attached to the ceiling of the polishing room, 30

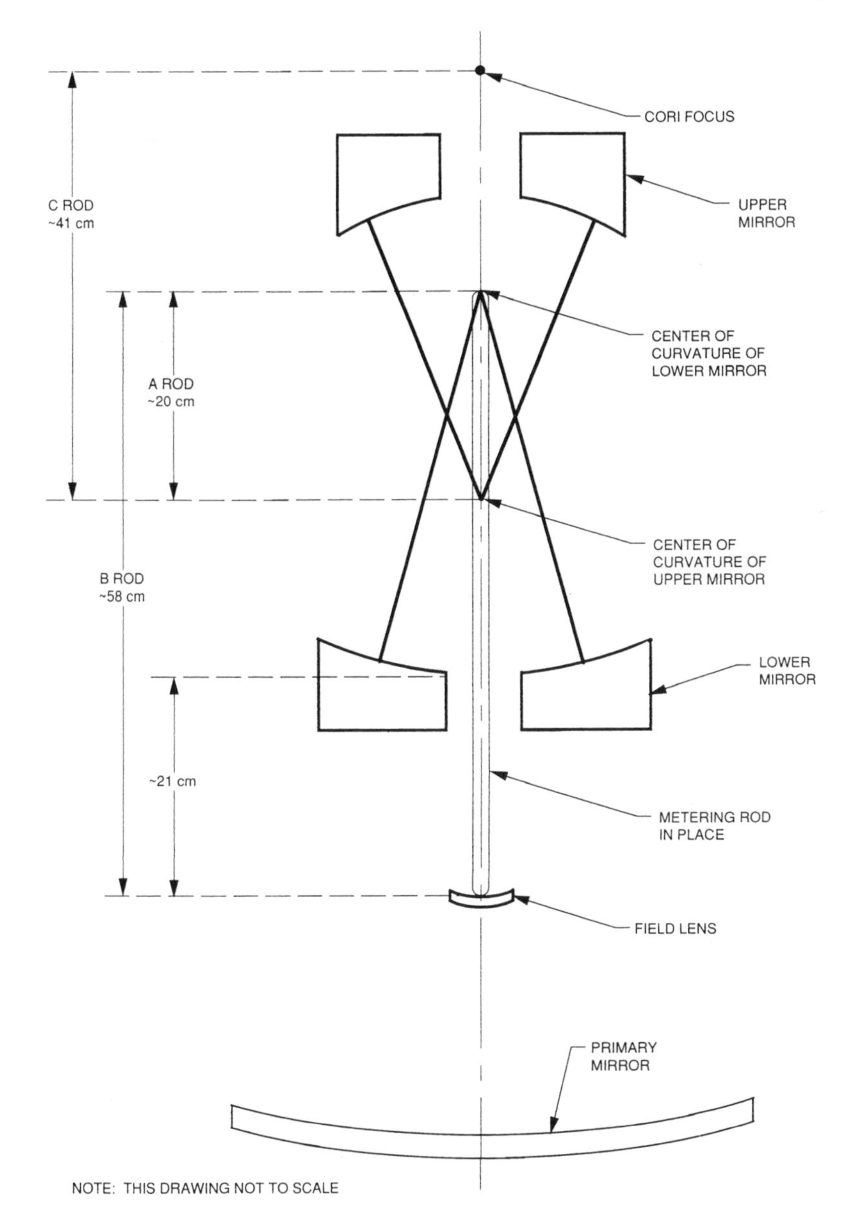

Fig. 4.3. This diagram, not to scale, shows of the arrangement of mirror and lenses inside the null corrector, as well as the measuring rods used to position them relative to each other. (Photo courtesy of NASA)

feet above the floor. To measure the shape of the telescope's mirror after each polishing session, the mirror would be slid on tracks into the polishing room and positioned under the small ceiling room so that the null corrector could look down at it. The lens on the null corrector that needed repositioning was on the bottom of the corrector (looking through the small room's floor to the telescope's primary mirror) and therefore required scaffolding from below to reach it.

It had taken several weeks to get the null corrector's two mirrors precisely placed. Now they only had a few days to get the lens in position.

It is important to recognize the precise nature of the distances involved. The tolerances were unforgiving. What made placement even more complicated was that the lens had to be positioned an exact distance not from a mirror but from a point in the air in which the mirror's reflected light came to a focus. To find this spot, the company had manufactured a measuring rod (the B rod in figure 4.3) of the exact length and made of Invar, a material that did not shrink or expand because of temperature changes.

Furthermore, the measurement had to be done against the exact center of the rod's end, aligned precisely with the centers of the lens and mirror. To prevent a misalignment, the engineers put a nonreflective cap on the end of the rod with a small hole in its center, thereby only exposing its center. They then fit the rod into position, using a laser beam to get its far end aligned precisely with the focal point of the mirror. If the laser beam missed the hole in the cap, it would not reflect, and they would know that the rod had to be repositioned.

Once the rod was in position, they then began lowering the lens to bring it in alignment with the other end of the measuring rod. For some reason that no one understood, however, the screws that moved the lens could not lower it enough. The rod's bottom end still stuck out below the lens.

It was puzzling, but there was no time to dwell on minor puzzles. The project was seriously behind schedule and had no money for intellectual exercises. The rod indicated that the lens needed to be lowered another 1.3 millimeters. Charles Robbert, who was one of the technicians trying to get the lens into position, remembered how the crew scrounged up three household washers, flattened them slightly to make them work, then fit them in the bracket holding the lens so that the adjusting screws could now lower the lens enough to place it in line with the end of the measuring rod.

They then removed the rod, removed the scaffolding, and declared the null corrector ready for use, carefully installed in its tower room where it would be difficult and unlikely for anyone to inspect its assembly.

No one from NASA was there to check what they were doing, since at the time the project's managers and scientists were too busy fighting over the budget issues surrounding Black Saturday. And even if they had wanted to be there, because of the military work that Perkin-Elmer was doing, the security at the Danbury plant kept the number of NASA observers to a minimum. Thus, the two science working group scientists who were supposed to verify Perkin-Elmer's telescope work, Bill Fastie and Dan Schroeder, were not present; in fact, they were never told what happened. Even the man who had designed the reflective null corrector, Abe Offner, was not advised that extra washers had been added. Any one of these people would have considered the 1.3-millimeter discrepancy worrisome and would have wanted to take a very careful look at the null corrector before using it.

On August 3, only eight days after Black Saturday, Rigby's crew began the fine polishing. For the next nine months they would methodically polish the mirror for a few hours or days, then move it on tracks from the polishing machine to below the null corrector, take a measurement to see where they needed to do more polishing, then move it back to start the whole process over. For reasons that no one understood, the null corrector was telling them to grind off a lot more material from the outer regions of the primary mirror than expected.[64] No one cared to question this situation too deeply, however. Time was money, and they didn't have a lot of either. The grinding went on, inexorably.

■ ■ ■

At the root of this tragic situation was the decision by James Fletcher in 1974 to limit the project's budget to approximately $300 million. This decision was not something that the scientists, contractors, and bureaucrats in and out of NASA had to accept. They could have told Fletcher that the telescope they could build at that price wasn't worth building, and either accepted its cancellation or forced Fletcher to back down and get them more money. Instead, they accepted his number, knowing that it was plainly unrealistic.

Moreover, no one either in or out of NASA was forced to accept forever the byzantine management structure that Jesse Mitchell and the management at NASA imposed on the project in the 1970s. When it became obvious in 1980 that the inherent conflicts between Marshall, Goddard, Lockheed, and Perkin-Elmer were costing money and making it impossible to build the telescope efficiently, someone could have stepped forward and said, "Enough!" Instead, everyone made believe they could make it work, and looked the other way.

Nor did the 1980 Black Saturday uproar help to fix these problems. Instead, there followed an endless series of new crises and budgetary panics. In July 1981 the managers at Perkin-Elmer, still short of money and resources, came to Speer and asked him for help. Speer said no. P-E, as the company was often called by people in the industry, was to put its nose to the grindstone and barrel on.[65]

Unable to get the job done, the company's managers simply focused all effort on the chore at hand, letting everything else fall by the wayside so that they fell even further behind schedule each week. By the middle of 1982 the situation had become so desperate that Perkin-Elmer's managers once again arranged a meeting with NASA, this time with James Beggs, then administrator.* This time the agency was more generous, giving P-E permission to hire two hundred more people, on the condition that it reorganize the project's management. P-E hired a new telescope manager, Don Fordyce, who had previously worked on the space telescope at Goddard and was now a manager in the private sector. With these changes, both Beggs and Speer thought the situation at Perkin-Elmer was finally under control.

It wasn't. Visits to Perkin-Elmer over the next few months by Speer and others from either Marshall or headquarters showed little real improvement. At the quarterly science working group meeting at the end of October, Speer had to admit to the scientists that P-E was fourteen weeks behind schedule. That same day Speer noted in a letter to P-E manager John Rehnberg that "the rapid erosion of the primary mirror

* Beggs had replaced Robert Frosch in July 1981, following the change of administrations from Carter to Reagan on January 20, 1981.

assembly schedule, the lack of a schedule incorporating the focal plane structure modification, and the manpower and cost increases over budget, indicate to me that a realistic plan leading to delivery of the [optical telescope] does not exist."[66]

On November 2, 1982, Speer made a report to Beggs outlining these problems. In that meeting, however, the sense that Speer tried to convey and that was accepted by almost everyone was that though things were tough, the situation was still under control. To Speer, if they could get Perkin-Elmer fixed, their problems would be over.[67]

To Don Fordyce at P-E, however, things were a mess. He had spent the last five months reviewing the project's status, and had come to the conclusion that things were unfixable. "They were in a hell of a lot more trouble than I thought," he said. "It was very obvious that we were not going to meet the schedule that was there with the manpower we had."[68]

Just before Christmas he finally decided that without some serious financial help Perkin-Elmer simply would not be able to finish the job, and to get it he needed to go as high in the chain of command as possible. At a January 14, 1983, briefing at NASA headquarters he laid out the seriousness of the impending disaster to Sam Keller, who had taken over for Frank Martin as the deputy associate administrator for space science at NASA headquarters. Fordyce's story was exactly the information Keller needed. He had been worried about the management of the space telescope project and had wanted to gain some control over it. He went to administrator Jim Beggs and proposed an increase in funds and a complete reorganization, with Keller in charge.[69]

For Beggs the situation was incredibly embarrassing. The annual NASA budget had already been approved by the White House and was printed, ready for its public release in mid-January. To now announce that the budget was wrong was awkward at best; at worst, it made Beggs look like he had no idea what was going on in his agency.

To his credit, Beggs took the heat, and negotiated with Congress what government bureaucrats call a "reprogramming," a process whereby they rearrange money within their budget, transferring it from one project to another. Suddenly, the space telescope had a gigantic infusion of cash. Hundreds of engineers were hired by Marshall, and put in place at Perkin-Elmer to help them get the job done. For the first time since it was approved in 1977, the project had the resources it really needed to get built. In fact, the buy-in that elected officials like Shipley and Proxmire had dreaded had occurred. Rather than cancel the telescope and have the money spent go to waste, Congress was prepared to accept whatever it took to get it launched and in orbit.

Beggs wasn't finished reorganizing the project, however. In order to save his own face, he needed a scapegoat. So did Bill Lucas, the director of the Marshall Space Flight Center. If Lucas couldn't give Beggs a sacrifice, Beggs might very well have the ax fall on him. In a letter to Beggs in April Lucas announced that he was instituting major changes in management, and had replaced Fred Speer as manager.[70]

The irony of this decision is that of anyone in management at Marshall, Fred Speer was probably the most conscientious. He might have not solved the telescope's budget problems, but it wasn't for lack of trying or for looking the other way.

Lucas, however, was far different. In fact, his methods of operation often acted to encourage people to look the other way. For example, he would hold quarterly meetings at Marshall where each project was supposed to describe its status. Everyone understood, however, that they were forbidden to discuss any problems during official presentations. There were even cases of individuals getting called on the carpet for mentioning areas of concern during their talks. "[The quarterlies are] highly choreographed," said John Brandt. "You're supposed to submit anything that you want to show weeks in advance so that it can be censored." Thus, in order to find out what was really going on people would instead go out into the hallway where the conversation was private.[71]

Unfortunately, there is no guarantee that government affairs will be just or fair. Speer was removed, and the project moved on.

■ ■ ■

Even so, the pattern of looking the other way did not stop. Maybe the worst moment came on May 26, 1981, when optical engineer Albert Slomba did a test to find the center point in the curvature of the primary mirror. The *refractive* null corrector he used for this test was not as accurate as the *reflective* null corrector that had been used to measure the mirror's shape while polishing. It was a small device that was attached

Fig. 4.4. Jim Westphal, Ed Weiler, Richard Harms, and Bob O'Dell express their thoughts on the quarterly meetings at Marshall. From right to left: "See no evil [O'Dell], hear no evil [Harms], speak no evil [Weiler], do no evil [Westphal]." (Photo courtesy of Ed Weiler)

under the big null corrector and swung into position below it so that a laser could be shone through it.

Slomba's goal was to locate the bottom of the primary mirror's curve, thereby confirming that the polishing had placed it in the center of the mirror. Slomba's test was not intended to measure the curve of the mirror. That had been done with Lucien Montagnino's big reflective null corrector. To his gratification, his test showed him that the bottom was exactly where it should be. In the middle of his image the fringes were straight parallel lines, as expected.

However, these central parallel fringes did not remain that way. On the outer edges the fringes became bent, as if the outside curvature of the mirror was not what it was supposed to be.

"This is a little strange," he thought as he looked at the fringes. From everything Slomba knew about optics, those curving fringes indicated there was an error in the mirror's curvature, something that at this stage of the project shouldn't be there. He picked up his two-foot-square photographs and walked over to Lucien Montagnino's office to talk to him about it.

As they pondered Slomba's interferograms, Montagnino's confidence in his big null corrector eased Slomba's mind. Slomba's small null corrector was far less accurate, producing grainy and messy images. If it had been the only tool available to measure the shape of the telescope's mirror, they would never have attempted to make the mirror at all. Eventually Slomba left to write his report, satisfied that if Montagnino wasn't worried he shouldn't be either. He decided to look the other way.

So did everyone at NASA when Slomba's report was passed on to them. The outside curved areas of his interferogram were dismissed for the same reasons expressed by Montagnino. Slomba's instrument was simply too crude to precisely measure the curvature of the primary mirror.[72]

Nor was Slomba's test the last time people at NASA or Perkin-Elmer looked the other way. Late in the polishing process, at about the same time Slomba did his test, a group of senior Perkin-Elmer opticians did a technical review of the project. One of their recommendations was to do a second test of the mirror's curvature, independent of Montagnino's reflective null corrector. As they noted in their review, "The purpose of this test would be to uncover some gross error such as an incorrect null corrector." This report was given to Jack Rehnberg, the company's manager in charge of the space telescope project. He did nothing, saying in later years that he did not remember seeing the report.

The essence of his actions, however, was a decision to look the other way. Making another test at this stage was expensive, and the risk of finding a big error was something the company simply couldn't afford. Better to do nothing and hope for the best.

Nor did the pattern of looking the other way end here. Carl Fuller, the man NASA assigned to check the quality of the mirror, was not an optician, but was never given any budget to bring in experts to check Perkin-Elmer's work. When it came time for him to sign off on the work, he refused. Unable to satisfy himself that the mirror was correctly made, he was unwilling to put his name to it. Yet, no one at NASA seemed bothered by this. Similarly, Danny Johnston, NASA's resident engineer at Perkin-Elmer, eventually received a report describing how washers were added to make the null corrector work. Rather than forward it to anyone at Marshall, including Bob O'Dell, he put it in his drawer and forgot about it.[73]

Nor was the space telescope project the only place at NASA where people were willing to look the other way. All throughout 1985 the engineers from the factory that built the shuttle's solid rockets had been warning both their managers and NASA officials from Marshall that cold winter temperatures could have a negative effect on the O-rings that sealed the gaps between the rocket's segments. As Morton-Thiokol engineer Roger Boisjoly wrote in a memo in July 1985, "It is my honest and very real fear that if we do not take immediate action to dedicate a team to solve the problem . . . then we stand in jeopardy of losing a flight along with all the launch pad facilities."[74]

Like Montagnino, like Slomba, like everyone else on the space telescope project, the managers at both Marshall and Morton-Thiokol looked the other way. On January 28, 1986, the space shuttle *Challenger* was ripped apart 74 seconds after launch.

For the Hubble Space Telescope, officially given that name in 1983 to honor American astronomer Edwin Hubble, the loss of *Challenger* was significant, delaying its launch an additional four years. For the project the delay was actually helpful, as it gave NASA, Lockheed, and the Space Telescope Science Institute the time to clean up a lot of the loose ends. For example, soon after *Challenger* they realized that the solar panels produced less power than anticipated and needed a fix. "We'd have launched on time," said chief engineer Jean Olivier. "We'd have had a much more sporty mission out of it; we would have had to control the thing much more carefully."[75]

Nonetheless, the delay once again put humanity's first sharp look at the heavens on standby. When the telescope was finally lifted into space by the space shuttle *Discovery* on April 24, 1990, it was almost twenty years since its design had begun and forty-four years since Lyman Spitzer had first proposed it.

At last, however, after eons of peering at the sky through foggy glasses and seeing only a blur, the human race was about to get its first unblunted view of heaven.

At least, that's what everyone thought.

5

Saving It

It was one of those priceless rare moments that everyone in the world of science dreams of. After decades of work and centuries of waiting, the first image from an optical telescope above the atmosphere was about to appear on a computer screen.

Chris Burrows leaned over the shoulder of the console operator at the Space Telescope Science Institute's control room to get a close look. Originally a high-energy physicist, Burrows had come to the institute in 1984 after working in Europe in industry as well as on the European Space Agency's Hipparcos astronomy probe. His father had been a scientist at Goddard when he was growing up, and taking a job at the institute was his chance not only to come to the back to the United States but also to get in on the ground floor with the Hubble Space Telescope.

Suddenly, that first image popped onto the screen, and a gasp of relief and excitement percolated through the room. The screen showed a scattering of faint stars, including a single bright star off to one side. The console operator played with the contrast, making that star's central core stand out. "Everybody was very, very pleased," Burrows remembered a year later. "The immediate reaction was wow! Look at that core."

And yet, not everyone was so happy. Roger Lynds was puzzled. A member of Westphal's Wide Field/Planetary Camera team, Lynds was an experienced astronomer working at the National Optical Astronomy Observatories in Tucson, Arizona. To him, there was something about the way the light formed around that single bright star that did not look right. In fact, he had never seen a "first light" star look anything like this—unless there was something badly wrong with the telescope's optics.[1]

■ ■ ■

The pressure to shoot this first image from Hubble had been indescribable. Despite careful scheduling and a deliberative plan for releasing Hubble's first images, everyone even slightly aware of the telescope's existence had wanted that first image *now*.

The plan, worked out by engineers at Marshall and Hughes-Danbury Optical Systems (which was what Perkin-Elmer was renamed after it had been purchased in 1989 by the Hughes Aircraft Company, a division of General Motors), had been to devote the first six weeks in orbit doing what the engineers had labeled "orbital verification." During this time period they would carefully get Hubble's mirrors aligned and focused. In addition, it was necessary to give the telescope time to vent any leftover gases and cool down to the proper temperature before turning on certain instruments. The Faint Object Camera, for example, operated at very high electrical voltages, which could arc and destroy the instrument if there were any residual gases available.[2]

Then, after this engineering work was completed and the engineers at Marshall were satisfied that the telescope was functioning properly, they would step back and, after almost twenty years of being the telescope's lead center, turn over Hubble's operations to the engineers at Goddard and the scientists at the institute.

Even then, it would not be time for unbridled scientific research. Now the science instruments would have to be tested and certified operational. Before launch it was assumed that this "science verification" phase would also last about six weeks, followed then by a full scientific program. If all went well, Hubble would begin taking its first images sometime within the first month after launch, during this verification process.

Nothing went well, however, almost from the moment astronaut Steve Hawley used the shuttle's robot arm to lift Hubble from *Discovery*'s cargo bay and send it out on its own. First, controllers needed two attempts to get the telescope to open its aperture door. The first try was aborted when the telescope went into safe mode, shutting itself down because the computer sensed problems with Hubble's high gain antenna.

Then, ground controllers found it difficult to lock the telescope onto any target for more than a few minutes. For reasons that were not at first clear, the telescope seemed to shake and twitch far more than its guidance system could handle. Eventually the source of this jitter was traced to Hubble's solar panels. Built by the European Space Agency as their part in the project,* the panels were simply not rigid enough to handle the temperature changes that occurred as the telescope orbited from night to day and back again. Each time Hubble passed through a sunrise or sunset, the panels would begin to shudder as they changed temperature, which in turn made the telescope shake for several minutes. Since the time between sunrise and sunset was only about forty-five minutes, and it took the vibrations from four to thirty minutes to settle down after each transition, Hubble was left with very little total stable viewing time, far too brief for the kind of long exposures astronomers needed to see faint objects in deep space.[3]

At the same time, the Marshall and Hughes-Danbury engineers were having trouble getting the telescope aligned and focused. The plan had been to use three white light interferometers that were built into the fine guidance sensors to focus the telescope. Called optical control sensors, these worked somewhat like the null correctors that had been used to measure the mirror's shape during polishing. The telescope would be aimed at a star, and its light would produce fringes after reflecting off the primary and secondary mirrors. By adjusting the tilt and position of the secondary mirror until the fringes became straight lines, the engineers would get Hubble focused quickly and far more precisely than if they tried to do it by taking pictures.

Unfortunately, the data being produced by these sensors made no sense. Instead of one or two fringes, they were producing so many that the engineers weren't even aware of them. The fringes were squeezed so closely together that all they saw was a blur. "It looked like garbage," remembered Pete Stockman, who was then deputy administrator of the institute.[4]

By the middle of May, almost a month after launch, the telescope seemed no closer to doing science than it had on the day it reached orbit.

* As part of the political negotiations with Congress to fund Hubble, Congress had insisted that NASA include international participation. The eventual agreement between NASA and the European Space Agency had the Europeans building the telescope's solar panels as well as one science instrument, the Faint Object Camera, in exchange for 15 percent of all observation time.

Complicating this situation was the pressure to take and release some pretty pictures. For decades ground-based telescopes had repeatedly made a big deal about what they called their "first light," the moment when the telescope was aimed at the sky and light would be allowed to travel through its entire system for the first time until it reached the eyepiece where the astronomer could see it.

Consider the hoopla surrounding the 200-inch Hale Telescope's first light. The June 3, 1948, dedication ceremony on top of Palomar Mountain in California was a truly exciting event, anticipated for decades. As a result, it included a thousand invited guests, speeches both enthusiastic and pretentious ("It will not be this telescope and all that it symbolizes that have led [humanity] to the doorstep of doom; it will be the impotence and immaturity of his moral values"), press tours, and demonstrations showing the rotation of both the 1000-ton telescope and its 137-foot-diameter dome. When the giant telescope was finally made available to the press the next night, reporters were dazzled with a look at Saturn and its rings, producing front-page news for days running.[5]

Though more than forty years had elapsed, the memory of that exciting moment still resonated with the public.

Unfortunately, though many people at NASA or among the scientists recognized that the country was excited about Hubble, none of them realized the depth of that fervor. Whenever the issue of public images came up at science working group meetings, it producing little interest or concrete planning. The scientists had personal reasons for resisting or not caring much about these public images. On one hand, their research was often so focused and esoteric that the pretty pictures were irrelevant, and did nothing to answer the questions they were asking. On the other hand, the public images might work so well at answering their questions that the images could become a threat, both to their scientific careers and to their personal research. Men like John Bahcall, James Westphal, John Brandt, and Robert Bless had dedicated years of their lives building Hubble. In exchange for that sacrifice they had been given guaranteed time on the telescope, ahead of anyone else. The use and proprietary control of those data was their "pay" for all that labor. Provide those data to everyone in the world and it might be possible for some outside scientist, who had done nothing to get Hubble built and therefore had made no sacrifices, to do an instant analysis and beat them to publication.

In fact, this very thing had happened only a few years earlier when *Voyager 1* had made its fly-by of Jupiter. Its first pictures of the moon of Io had revealed a violently volcanic surface, including one image of a plume, the first time an active nonterrestrial volcano had ever been spotted. Because these images had been released to the public the instant they arrived from space, it had been possible for any scientists to put forth their own interpretation of the data, even if they weren't part of the spacecraft's science team. Most outside researchers who considered doing this first discussed it with the science team that ran *Voyager 1*, and backed off when it was pointed out to them that the public data were incomplete and that the science team was really in a better position to analyze it.

One postdoctoral student who was not part of the science team did not back off, however. Encouraged by his thesis advisor ("a politically prominent planetary scientist"), he wrote up his interpretation of Io's volcanism and got it published in *Science*. When Brad Smith, head of *Voyager 1*'s imaging team, questioned the advisor about this action, the advisor told him "(1) that anything that gets into the newspapers is public property, (2) that this sort of thing is done all the time and (3) that he does not believe in proprietary rights to spacecraft data anyway."[6]

Because of these memories, the members of Hubble's science working group had negotiated long and hard, both among themselves and with NASA, to make sure their rights would be protected. John Bahcall more than anyone took the lead in working out the deal. First their guaranteed time was set, totaling 350 hours per instrument team, arranged as a sliding scale over time, with the guaranteed time observers initially getting 100 percent of the available U.S. time during the first seven months.[7] They then had to make sure their own research ambitions did not conflict. Once again the negotiations were initiated by John Bahcall. These issues were hammered out during a particularly passionate meeting, which he chaired, at the Institute for Advanced Study in Princeton in 1985. Since many of the scientists wanted to look at the same objects, first rights to each object had to be determined. They decided that different observers could look at the same star or galaxy or nebula if their fundamental research goals were different or they were using different science instruments (such as a spectrograph versus a camera). If their goals or instruments were the same, however, the targets had to divvied up so that no one stepped on anyone else's toes. Each instrument team made up a list of their primary targets, and spent the meeting haggling over their picks, almost like kids trading baseball cards. "The meeting in fact turned into a love feast," remembered Jim Westphal. "There was all manner of real effort by literally everybody who was there to work with each other, to cut deals between each other, to share targets, and to be more efficient."[8]

Nonetheless, the delicacy and complexity of these negotiations left little room for the scientists building Hubble to think deeply or carefully about the problem of providing images for the public. Even as late as mid-January 1986—just prior to the *Challenger* accident, when Hubble's launch was expected in only ten months—the science working group had devoted almost no time at its meeting to the issue. And when the subject was raised at the April 1986 meeting, it was scheduled for the very end of the meeting and garnered little debate from the members. Frustrated by this lack of interest, Riccardo Giacconi, director of the institute since 1981, proclaimed that if they didn't take some pretty pictures to show the public, "we are crazy!"[9]

Despite this plea, the guaranteed time observers remained either uninterested or defensive whenever anyone from the public affairs offices of either NASA or the institute suggested using Hubble to make early observations of any well-known or significant astronomical object. By 1989, the institute had its own public outreach department, and its head, Eric Chaisson, was lobbying to image and release a few of what he called "pretties" as soon as possible after launch. At first he proposed imaging objects not reserved by the guaranteed time observers. However, because these proposed observations had not been approved by the telescope's time allocation committee, they were rejected by the scientists. From their perspective, why should Chaisson bypass this committee when they couldn't?

Chaisson then decided to select a variety of spectacular objects already reserved by the guaranteed time observers, and ask the observers to approve their release.

This idea was also shot down. In fact, the issue became so heated that John Bahcall actually threatened Eric Chaisson in public.

Though Bahcall mostly intended to use Hubble to do an optical survey of quasars in an effort to figure out the nature of these distant and powerful objects, he also planned to take the first good pictures of a number of well-known and spectacular globular clusters, including M15 in Pegasus, described by one astronomer as "among the dozen finest objects in the northern sky."[10]

When Chaisson presented his list of proposed observations at the fall 1989 science working group meeting, most of the scientists at the meeting saw this proposal as a threat to the work they had done for the past two decades. For example, Westphal noted that if one of his objects ended up in the newspaper, "anyone could put a ruler to it and then go out and publish a paper." He would be scooped, just as had happened with *Voyager 1*. Bahcall in turn challenged Chaisson on his access rights to M15, telling him that since he was not a guaranteed time observer he had no right to release images of this object to the public.

"Eric had a way of inciting people's worst tendencies," explained Ed Weiler, who had taken Nancy Roman's job as NASA's chief astronomer when she retired. Weiler also noted that Chaisson was new to the project. "Unlike Eric Chaisson, most of these scientists like John Bahcall had worked on Hubble their entire careers. So this young whippersnapper coming in and telling them how they were going to do things did not rub some people very well."

Chaisson persisted. The debate became more heated. Finally, Bahcall pointed at Chaisson and said quite bluntly, "If you look at those objects before I do, I'll kill you."

Eventually the scientists won out, and as a result, no public relations program was organized. Instead, images would only be released when the scientists agreed to let them be released. "There was absolutely no plan," said Ray Villard, who was Chaisson's deputy at the time. Villard remembered one scientist saying, "They [the public] can afford to wait."[11]

Not surprisingly, having no plan was not going to go over well with either the public *or* the press. The day before launch, NASA held its standard pre-launch press conference, featuring representatives from the shuttle program as well as the space science program. Lennard Fisk was now science head, the same job that Homer Newell, John Naugle, and Noel Hinners had held before him. When asked about first light, Fisk was unprepared to answer. From his perspective, that first light image was nothing more than an engineering test, an incredibly boring handful of stars, no more. He, like the scientists, hadn't thought anyone would be interested in seeing it.

The reporters thought differently. They had become used to decades of on-the-scene NASA history-making, and expected to see that first light image, as it happened. When Fisk failed to give a clear and straightforward answer to the first reporter's question, several other reporters in the room followed up, peppering Fisk repeatedly about that first light picture. What was NASA's strategy for releasing that first image? Who is going to decide when to release it? What will be the telescope's first target? Are you worried that the poor quality of that first image might disappoint the American people and hurt your budget negotiations with Congress?

At first, Fisk tried to improvise, noting how after the telescope was aligned NASA had planned to make public the first good scientific result, illustrating and comparing Hubble's resolution with a comparable ground-based telescope. "We were simply looking to have a way in which this, you know, proceeded in an orderly fashion."

The hail of questions quickly changed Fisk's mind, however. Before the press conference was over he had unequivocally agreed to make that first light image available as it happened, no matter how pedestrian it looked. "If you want to see the bits, come see the bits," he added.[12]

Because of the telescope's initial operational difficulties as well as the puzzling fringes produced by the white light interferometers, however, first light didn't happen the first week after launch. Nor did it happen the second week. Or the third week. Only after almost four weeks in orbit did the engineers feel comfortable enough with their tracking and focus woes to risk taking such an image.

With much fanfare, the day for first light finally came. Charles Pellerin, who worked under Fisk as head of NASA's astrophysics division (the same job that Jesse Mitchell, Bland Norris, and Frank Martin had held in past years), announced at a Sunday press conference on May 20 what was going to happen. "At 11:12 this morning the shutter opened and the Wide Field/Planetary Camera took its first exposure." Overall, the telescope had taken four exposures of different lengths, which, once downlinked and processed at Goddard, the public would see at the same moment they appeared to the scientists.

Once again, Pellerin noted NASA's public relations concerns. "If I had my choice this is not be the way we would do these things. We'd prefer to have a scientific result in front of us and call a press conference to come look at it but now you are involved in the middle of the process with us."

Jean Olivier, the project's chief engineer and the man who had babysat Hubble's construction from its inception, added, "Today is an engineering, if you will, evaluation." He further warned that the telescope's focus might not yet be ideal: "We are about maybe one fourth or one third of the way through this whole alignment process. It will be quite some time before we achieve the final product."[13]

In truth, the whole ceremony surrounding "first light" was somewhat absurd. For more than an hour the television cameras at Goddard showed nothing more than a handful of controllers sitting in front of their computer terminals, pressing their keyboards periodically, and generally not doing much. Once in a while a green indicator light on a computer would change to red, and an excited operations person would get on air to explain that another image had been downloaded.

Finally, the first image popped up on a computer screen. As expected, it was a boring image of a handful of stars, supposedly in a galactic cluster in the constellation of Carina in the southern hemisphere. The scientists sitting at the operations console, however, were not so sure. They spent most of the next hour mulling over the image, trying to identify the stars.

What made the situation even more absurd was that when the so-called first light image was unveiled at the Goddard control center it was not even the first time that image had been seen by humans. Prior to its appearance at Goddard it had already been seen and mulled over by the scientists at the Space Telescope Science Institute in Baltimore, including Chris Burrows and Roger Lynds. There, the mood was generally good. Hubble's incessant jitter and focus problems had caused many people to worry that the first image wouldn't show anything at all. Instead, they now had real data, with real stars, and could use this solid information to finish the telescope's alignment.

Roger Lynds, however, was concerned. The next day, at a crowded Monday operations meeting at Goddard, he remarked that the circular cloud of light surrounding that one bright star could only be caused by something wrong with the telescope's optics, something called spherical aberration. The Goddard operations manager who was running the meeting, Joe Ryan, immediately told Lynds to stop being an alarmist. The situation with the press was crazy enough without starting new and wild rumors.[14]

Lynds was not the only person puzzled, however. Chris Burrows had immediately begun an analysis of the image and was equally confused. Burrows's main expertise at the institute was to understand the minute eccentricities of the telescope's optics, and then apply that knowledge to help scientists who used Hubble interpret their data. Prior to launch he had written the telescope's main handbook, describing in precise and microscopic detail "how [Hubble's] design affects the images produced at the science instruments."[15] In addition, he and Hashima Hasan (who was another scientist at the institute) had written software to simulate the kind of images that they expected Hubble to produce. Now the telescope was in orbit, and it was necessary for Burrows to compare those pre-launch predictions to the actual data. The same day Lynds expressed his worries in public, Burrows fired up this software and began comparing the actual image with his various simulations.

The brightest star in the image, Iota Carina, was luminous enough for him to analyze. The trouble was that it didn't look anything like he had predicted. Instead of 70 percent of its light focused in a tiny spot a tenth of an arc-second across, most of that light was spread out, with only 10 percent in the central core with the rest forming a hazy and fuzzy cloud ten times wider. What made it even more puzzling was that within that fuzzy cloud were strange "tendrils" pointing out from the core.

Making the problem even more baffling were the number of variables that could explain what he saw. In order to sort through them all Burrows spent the night writing a new program, including within it anything that could cause fuzziness in an optical system. Depending on the values applied to each variable, the software would create an image. When the image matched what he saw he would know which variables were causing the fuzziness.

For two days he struggled with the data, sleeping very little and spending untold hours at work. "Often I would sleep in my office, napping and then waking up and going on with stuff." he remembered. "The whole period was somewhat of a blur."[16]

By Tuesday morning he was convinced that something was seriously wrong. Later that day, at the daily operations meeting at Goddard, he was so anxious to present his results that he interrupted a talk by Jim Westphal. "Look, I can fit these images," he announced eagerly.

Fig. 5.1. Chris Burrows in 1986, holding his 18-month-old son Richard. (Photo courtesy of Chris Burrows)

Westphal graciously stepped back. "Yes, let's hear what Chris has to say."

Burrows described that the star image could only be explained if there were two kinds of optical imperfections in the curve of the telescope's mirror: spherical aberration and coma. Draw a cross section cutting through the center of a curved mirror. It should look like a left-hand parenthesis, like so: (. Now draw a second line pointing out from the center of that curve, like so: (——. In a correctly shaped mirror, the curve of the mirror focuses all the light to a single point on that horizontal line. With spherical aberration, the mirror's curve causes light from different parts of the mirror's surface to focus at different points along that horizontal line, rather than at a single point. As a result, a star will look like a circular fuzzy blob. With coma, the mirror's curve brings all the light to a focus at the same distance above the mirror, but light from the mirror's outer edges is focused offset from that horizontal line rather than on it. The result is a blur that resembles a tiny comet with a wide cone-shaped tail.[17]

Burrows thought that the distortion in Iota Carina was mostly the result of coma, with some spherical aberration. He tentatively suggested at this May 22 meeting that they might have to consider using the twenty-four actuators attached to the back of Hubble's mirror to correct the problem. These mechanical pressure points were designed for making coarse adjustments to the mirror's shape.

At this time, however, no one was willing to believe such a conclusion. The Marshall and Hughes-Danbury engineers, led by Jean Olivier and Bob Basedow respectively, were still struggling to get focus using the optical control sensors. Until they had a better idea of why these sensors were not working as predicted, they were reluctant to attribute the soft star image to a fundamental error in the telescope's mirror.

Moreover, Burrows's analysis was incomplete. His theory of coma plus spherical aberration failed to explain everything in the star image. For example, he could not yet account for the strange tendrils and kidney-shaped shadows that emanated from stellar cores.

For the next week, while Burrows argued his case to no avail, Olivier and Basedow and the engineers under them continued to struggle with the optical control sensors, trying without success to use them to focus and align the secondary mirror. At the same time Jim Westphal was repeatedly calling for more pictures, saying that since the optical control sensors didn't seem to work, pictures were necessary to help find focus. After a week Olivier finally relented enough to schedule a second light image on May 31.

That same day a new person entered the picture. Sandra Faber had only joined Jim Westphal's Wide Field/Planetary Camera team (affectionately called "wiffpic" by team members after its four-letter acronym WF/PC) in 1985. A well-known and respected astronomer from the University of California in Santa Cruz, Faber was best known for discovering a link between the brightness of elliptical galaxies and the speed at which the stars within them moved. Later, she was part of a major research project that tracked the motions of hundreds of elliptical galaxies and discovered a gigantic general flow of galaxies toward a previously unknown density of matter, what astronomer Alan Dressler dubbed "The Great Attractor."[18]

Originally part of Spitzer's SEC Vidicon team—helping to write their proposal—Faber, like Lyman Spitzer, had been left on the sidelines when Westphal's CCD proposal won its bid to build Hubble's camera. However, as the years passed she was periodically hired by Westphal to do consulting work for him, and he found her to be his kind of person, sharp, honest, and straightforward. As he described her, she's "the quickest quick study I've ever been around." At the same time, James Gunn was looking for more scientists to add to the WF/PC team. He especially wanted someone to help Westphal with the team's management. After doing the prerequisite negotiations with their fellow team members as well as NASA's bureaucracy, Faber was added to the WF/PC team in 1985.[19]

She had missed first light because she had been doing observations of spiral galaxies using the 60-inch telescope on Palomar Mountain. Arriving on the morning of May 31, 1990, she not only got her first close look at the first light image, but also saw the second image taken that day of the galactic cluster NGC188. Located four degrees from the North Star, NGC188 is one of the most unusual galactic clusters known. While most galactic clusters are very young—sometimes only a million years old—NGC188's stellar population suggested an age exceeding 12 billion years, making it one of the oldest clusters known and more akin to the ancient globular clusters forming the halo of the Milky Way.

The May 31 Hubble image was not intended to study the enigma of NGC188's age, however. Instead, it was the first serious attempt to use WF/PC to focus Hubble's mirrors.

Fig. 5.2. Sandra Faber. (Photo by Tod Lauer)

Faber instantly noticed something strange about these two images. Depending on a star's location in the field of view, the shape of its radiating tendrils changed. To Faber, the most likely cause of these changes was interference from the supports holding the mirrors inside WF/PC, combined with some form of spherical aberration in Hubble's primary or secondary mirrors. By that evening she was arguing her case with Ed Groth, a long-time member of the science working group and the telescope's data and operations team leader, and Tod Lauer, one of Faber's postdoc students whom the team had hired. Neither were convinced.

All of them agreed, however, that if they could get a series of pictures taken by WF/PC at a range of focuses with the star placed at the center of the image they could very quickly define the cause of the image's fuzziness. According to this plan, the secondary mirror would be moved after each image, changing its distance from the primary and thus the focus.

The problem with this plan was that it wasn't the plan that the Marshall and Hughes-Danbury engineers had worked out for getting Hubble in focus. Olivier and Basedow remained reluctant to do it because it required moving the secondary mirror far more than planned. If the mirror got stuck during any of these focusing maneuvers the telescope would be left permanently out of focus.

For Faber, convincing them to change their approach was made more difficult by her newness to the project. The engineers didn't know who she was. She was further hampered by her lack of knowledge of the field of optics as well as her limited knowledge of both WF/PC and telescope. Compounding these obstacles were the strong disagreements between the scientists and engineers. The engineers at Hughes-Danbury, for example, were claiming that their software was able to fit the images and had found nothing wrong with the mirror. They refused to release these data, however, which made it impossible for anyone else to check it.

Meanwhile, Burrows was hopeful that the focus issues were caused mostly by coma. If so, they could fix the problem in two ways, first by tilting Hubble's secondary mirror slightly in just the right way and second, by using the actuators on the back of the primary mirror to change its shape slightly.

Though Faber was skeptical that coma was the problem, she was also eager to pursue any solution. She didn't know much about the actuators, however, and was having difficulty finding out. One time she spotted a Hughes-Danbury engineer in the hallway of the Goddard operations building. She thought, here was someone who would know exactly how the actuators worked as well as their limitations. She buttonholed him and began asking questions.

Even as he gave her sketchy answers to her questions, the man began walking away from her down the hall. "[He was] really brushing me off. We were walking down the hall and he was being rude in the way some people are deliberately rude. . . . They know you want to stop and talk to them but they don't stop; they just keep walking and they make you tag along. So I was tagging along, asking him about how I could learn about actuators and did he have any reports about them."

"What do you want to know about them for?" he asked as he moved away.

"Maybe there is something wrong with the primary," she explained. "We could use [the actuators] to fix it."

The engineer accelerated his walk. "You better not let anybody hear that you're talking about them," he said over his shoulder as he left Faber behind. "You'll get into trouble."[20]

For the next week, while she, Burrows, and her graduate student Jon Holtzman carefully picked apart the two images available to them and argued over the cause of the strange shapes they saw there, they also began a concerted campaign to get the project to take more WF/PC images. They wrote and presented new image proposals each day. Faber began inviting herself to the morning planning meeting, describing the work they were doing and pitching their proposals.

On June 8 Faber had to leave for about a week to do work for the Keck Telescope Science Steering Committee, of which she was then a co-chair. Before going she wrote a long memo to Jim Westphal, explaining the situation and the political landscape. According to this memo, Jean Olivier was "sounding increasingly desperate." She also noted how the Hughes-Danbury chief engineer, Bob Basedow, "is a smart guy, but he's fighting [the problem] with both hands behind his back." Because no one at his company seemed willing to admit there was anything wrong, he was unable to bring an objective mind to the problem.

She then explained what she had been trying to do. "Get the system to appreciate the need for a simple focus run," she wrote. "Look ahead and understand the [primary mirror] actuators and what they can do." Then she added, "I don't have any conclusive model yet myself. However, the situation I'm worrying about most is . . . large amounts of spherical aberration and some defocus. . . . I haven't told this to anyone else but Jon [Holtzman] and Roger [Lynds], but it keeps me awake at night."[21]

By the time she returned on June 17, Jean Olivier had had enough. After watching the Hughes-Danbury engineers struggle for another week with the optical sensors, he had finally agreed to the idea of doing a focus run using a series of WF/PC pictures, beginning on June 14.

These pictures only made the situation more depressing. The amount of fuzziness they showed suggested an error far greater than anyone expected. Holtzman pointed out something to Faber that shocked her further. One day, as they drove to Goddard to work, he explained how the units of measurement the project had been using to describe the focus problems tended to fool people into thinking that the error was not serious. The numbers were small and made the problem seem smaller. If the error was measured in a different way it made the grossness of the error much more obvious.

At that moment Faber had a flash of terrible insight. "Oh my God," she thought. "This is horrible."

Then, on Tuesday, June 19, Bob Basedow finally gave the project a detailed report on how the actuators worked, explaining that they could only change the primary mirror's shape by at most 10 percent, far too little to correct the now estimated error, regardless of whether it was coma or spherical aberration. Months later Faber scribbled an extra comment at this point in her daily notebook. "This is the moment we find out that we are doomed to failure!"

Still unable to convince anyone of her belief that spherical aberration was the primary cause of the focus problems, on Thursday, June 21, she and Holtzman decided to use Burrows's simulation software to create their own models. Holtzman fired it up and together they began to try to recreate the star images by using different amounts of spherical aberration. If they could create a simulated image that matched a real one they would prove that spherical aberration was the cause of the telescope's fuzziness.

All through Thursday, Friday, and Saturday, they labored over the computer, slowly calculating and printing out a set of six simulated images, each based on a different amount of spherical aberration. Periodically Faber would drive over to Goddard to pick up additional image data from Hubble itself. They would then lay their simulated printouts on a conference table next to the printouts of the actual images, comparing them.

By Saturday morning they were finally sure. One particular picture from Hubble showed a series of concentric rings surrounding the central star core, which they dubbed the "smoke-ring" image. One of their own simulated images duplicated this image almost precisely.

They knew that on Monday there would be the regular weekly project meeting, much more formal than the daily operations meetings and attended by all the most important players in the project. At that meeting they would have the chance to make their case. Moreover, they knew that the entire group of project scientists from all six science instruments were gathering in Baltimore for the already scheduled quarterly meeting of the science working group, set to begin on Wednesday. It was essential that the project had a clear idea about Hubble's status when that science working group meeting started.

They began to prepare their presentation for the Monday project meeting, with Holtzman creating viewgraphs showing their simulations side-by-side with the real images, and Bob Light, one of the other post-graduate students working on the WF/PC team, preparing viewgraphs showing the actual distribution of light produced by Hubble's image compared to its predicted performance.

Faber meanwhile kept going back and forth to Goddard to pick up more image data. As she walked into the control room late that Saturday afternoon, she found the engineers standing in a group, staring at a monitor showing the smoke-ring image and trying to figure out what could cause such an image. "They were completely mystified," she remembered.

She quickly sketched out a crude drawing, showing how spherical aberration combined with Hubble's optics could produce the smoke ring. "Wow," said scientist Keith Kalinowski. "You oughta go down and talk to those Hughes people." Basedow and his people had just been there and had been equally perplexed by the smoke ring.

Faber decided to take Kalinowski's suggestion, though "in one of my rare moments of tact I . . . sort of eased my way into the room." Bob Basedow and John Mangus, Goddard's optical expert, were there, discussing the situation. "I could see that people were very puzzled." Gently Faber offered to explain the smoke-ring image, once again making her crude sketch.

"Basedow's face went white," Faber remembered. "It finally dawned on him what was going on. The magnitude of it."

At that moment Mangus pulled from his pocket his own carefully rendered diagram, identical to Faber's but very professionally produced, to scale and labeled. He laid it on the table and quickly used it to describe the physics of the situation.

As Faber was leaving the room Mangus joined her. Out of curiosity she asked him, "John, you knew all about this. Why weren't you telling these people this? You had that sketch right in your pocket."

Mangus replied coyly, "I was waiting for you to show up." Though he had solved the problem enough to prepare his diagram in advance, he hadn't wanted to be the first to unveil it to the others. Like those who had been slapped down at Marshall for reporting bad news at quarterly meetings, he as a government employee feared the consequences of being the bearer of bad news.

At 1 pm on Monday, June 25, everyone gathered in Building One at Goddard to review the situation. Jean Olivier ran the meeting, first calling the engineers from Hughes-Danbury to lay out their thoughts on the project's status. Though Basedow in his talk finally agreed that the mirror had serious spherical aberration, he continued to describe it using the smaller unit of measurement that obscured the seriousness of the distortion.

Then it was time for Faber and Holtzman to speak. As much as she wanted to have the spotlight, Faber let Jon make the presentation alone. He was young, this was an ideal opportunity for him to showcase his talent, and she wanted to help promote her former student's future.

When he unveiled their viewgraph comparing the real Hubble image with their simulation, the room became hushed. For a few moments no one spoke.

Then Jean Olivier softly asked them to keep their results quiet until he could talk to people at NASA headquarters the next day. The debate had ended. No one could disagree. Hubble's mirror, intended to be the sharpest and most accurate mirror ever ground, had been ground wrong. The telescope was out of focus.

That evening Faber, Holtzman, Ed Groth, Bob Light, and a handful of others from the WF/PC team had dinner at a Japanese restaurant near their temporary offices at Bowie State University in Maryland. The

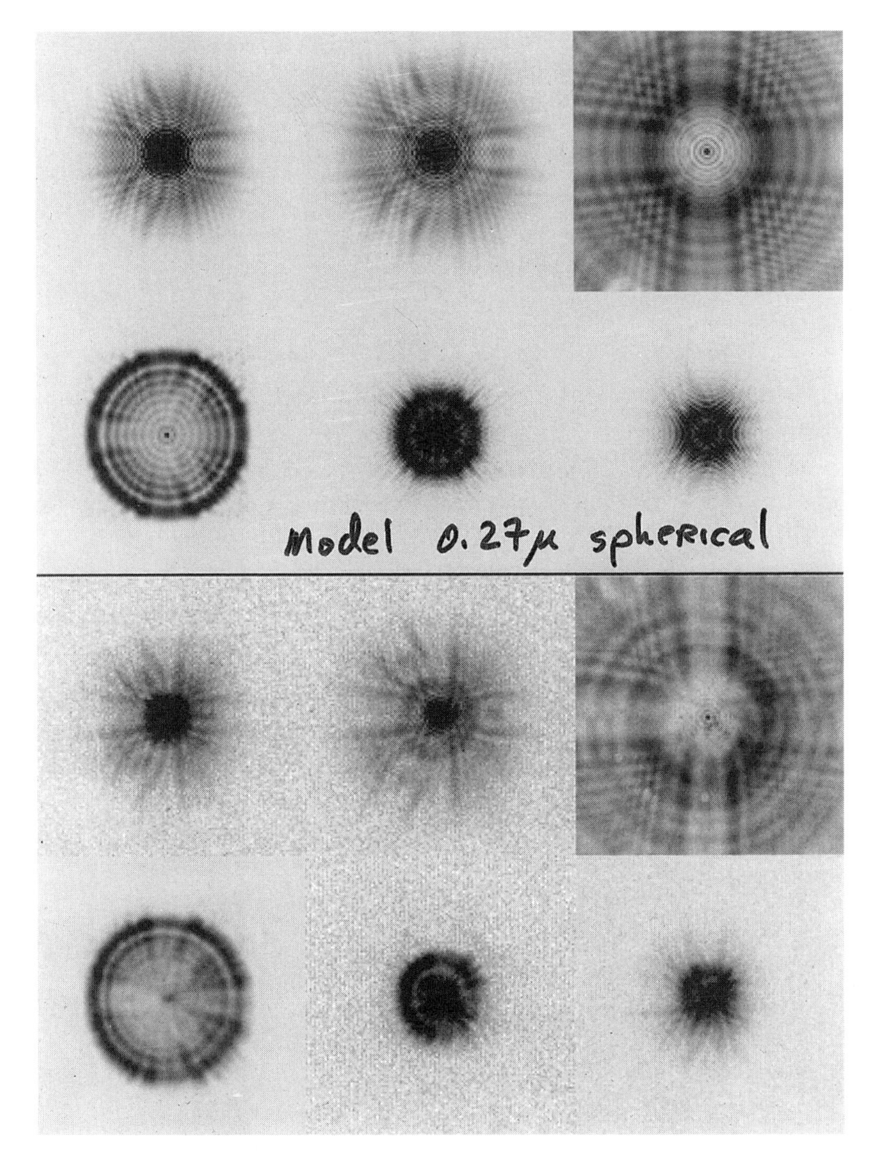

Fig. 5.3. The actual viewgraph used by Jon Holtzman during his presentation at the June 25, 1990, meeting, showing a comparison of simulated star images (top) with a set of real star images taken by Hubble (bottom). The "smoke-ring" image is on the lower left of both sets. (Viewgraph courtesy of Sandra Faber)

events of the past week had been for Faber and Holtzman an incredible roller-coaster ride. Simultaneously they had been both exhilarated and depressed by their conclusions. Unlike anyone else, they had successfully pinned down the problem to a remarkable accuracy. Moreover, they had done it in such a way as to convince everyone of their conclusion, despite the horrible consequences of that conclusion.

At the same time, they had been witness to what might be the greatest catastrophe to hit American astronomy ever. As they sat there and quietly ate sushi, Holtzman and Light, both young men at the start of their careers, worried whether Hubble's failure was going to destroy their future. Everyone at the table also speculated about the possibility of congressional investigations, the death of NASA, the cancellation of all big NASA projects, and the possibility of a backlash against big science. They even feared that, as Faber wrote in her notebook, "astronomy might become a pariah," with vicious "name-calling" between ground-based and space-based observers.

As they mulled over this grim future, they worried whether NASA would kill the second generation WF/PC-2 clone that had been under construction since the mid-1980s and was scheduled to be installed on Hubble during the first shuttle maintenance mission three years hence. Though some thought it might be canceled, Faber was hopeful. As she noted later, the clone was "costly, but could be a big, American-style rescue effort."

After dinner Faber had a teleconference with Westphal, Groth, and the project scientist Al Boggess. Boggess was very pessimistic, noting that the urge to kill Hubble at NASA headquarters was strong, considering the continual shortage of cash. Groth was even more glum. "My science is dead," he said sadly.

During that teleconference Faber got a side phone call from Tod Lauer, who suggested that since the aberration seemed so well defined, maybe computer software could clean it up. Faber passed the news to the others, suggesting that they needed to do simulations showing whether software could make Hubble useful again.

Two days later the scheduled quarterly meeting of the science working group began. After some cautionary and depressing preparatory remarks by Al Boggess, Jean Olivier carefully laid out the seriousness of the situation, noting that the telescope's capabilities were "severely compromised." As had happened two days earlier when Jon Holtzman had unveiled his viewgraph comparing the real image with his simulation, the room became very quiet.

After Faber and Burrows reviewed the data in detail, it came time for the various instrument scientists to assess the impact of the spherical aberration on their research.

Jim Westphal was first. The Wide Field/Planetary Camera was by far the most important instrument on Hubble, scheduled to use almost fifty percent of Hubble's telescope time for taking spectacular images. The consequence of spherical aberration for those images was therefore foremost on everyone's mind.

In the years since Westphal's JPL team had won the contract to build WF/PC, he had found himself at the center of a gigantic government science project, the last place he had ever wanted to be. Over those fourteen years his life had been consumed by it. He had reduced his scientific research, got himself relieved of all teaching duties, and reassigned everything else to other people at Caltech so he could devote his entire time to massaging the bureaucracy and doing what he considered the boring management work that was necessary to make any NASA project fly.[22]

Always at the end of all this tedium had been the dream of seeing the sky clearly, without obstruction; of taking neat and glorious pictures of some of the sky's most famous astronomical objects, of having "fun" using WF/PC to produce the first sharp pictures of the heavens ever seen by humanity. In outlining his planned use of his guaranteed time, Westphal had proposed looking at almost *anything* he could think of that would look interesting when imaged. Inside the Milky Way he had planned to image planetary nebulae like the Helix Nebula, T Tauri stars, the stars in the galaxy's bulge, globular clusters, and a collection of other weird galactic cloud formations. Outside the Milky Way he was going to aim Hubble at a number of peculiar and merging galaxies as well as the nuclei of several well-known nearby normal galaxies. He even planned to take a few very long exposures of blank parts of the sky, to see what he could find there.

Inside the solar system Westphal had hoped to take a close look at all of the planets as well as comets and asteroids. For example, when the two Voyager spacecraft had flown past Saturn in 1980 and 1981, they had seen what appeared to be dark spokes radiating out along the surface of the planet's rings. No one understood what these spokes were. Westphal hoped a Hubble picture would tell him.

Overall, his observations had had the chance to revolutionize astronomy.

Unfortunately, the Hubble's misshapen mirror demolished these plans. After spending fourteen years doing what he considered drudge work, Westphal, now sixty-one years old, had apparently nothing to show for it. When it came time for him to report on the consequences of the spherical aberration to WF/PC, he didn't even get up from his seat to go to the front of the room. He sat there, and in a short one-minute report bluntly summarized the disaster. "This essentially wipes out our entire science program."*[23]

Westphal's grief was not his alone. As each person stepped forward to give their report on the impact of Hubble's status on their instruments that June day the sense of tragedy and loss seemed to permeate the air. Duccio Macchetto, who was part of the European team that had built the Faint Object Camera, seconded Westphal's report. Only forty percent of the FOC's program was now doable, and even that portion would require three to five times the exposure time, something that the telescope simply couldn't afford. "Much pain," was how he summarized the situation.

Robert Bless talked about the High Speed Photometer, noting that while a good percentage of its work was still feasible, a significant portion either had been lost or was little better than what ground-based telescopes could do.

Richard Harms noted that about half the Faint Object Spectrograph's program was wiped out. The other half would take from two to five times longer to get the same amount of light.

* Less than a month later, Jim Gunn, the man who had convinced Jim Westphal to become PI of WF/PC, wrote an anguished email where he harangued both himself and the entire astronomical community about the failure of Hubble. "It was an ASTRONOMICAL FAILURE," he wrote, noting how the astronomers could only blame themselves for accepting NASA's way of doing business. "NASA's style, however, is killing/may have already killed us, and we will never get another chance to do anything about it, if indeed it is not already too late."

Gunn decided to get out. Soon thereafter he left as WF/PC deputy PI, a position he had held for a dozen years, and became the driving force behind the Sloan Digital Survey, which is making the first digital map of a quarter of the entire sky, with a catalogue of more than 100 million objects. (Gunn email, 7/11/90, Villard papers.)

John Bahcall was quietly explicit about the situation, noting that almost all his science using Hubble's cameras was now undoable. The only instruments that seemed to have any capability left for doing the kind of research they were built for were the spectrographs. They didn't need sharp images, and to get the amount of light planned they only had to lengthen the time necessary for each exposure. And since the cameras were no longer very useful, there was now spare time to devote to spectroscopy.

For Bahcall, the last few weeks had been very difficult. In February 1989 he had accepted the job of chairman for the astronomy and astrophysics survey for the 1990s, comparable to the one that Jesse Greenstein had chaired for the 1970s. He had just finished the report, and had faced a lot of heat from astronomers because of his decision, like Greenstein's, to force them to make choices and list priorities. At the same time his mother had had kidney failure, and he was scheduled to fly back to Louisiana the next day to see her and her doctors.[24]

"We should rename HST the Hubble Spectroscopic Telescope," he finished sadly. As important as spectroscopy was, Hubble was no longer the telescope to which he had devoted all these years of effort.

O'Dell, now a somewhat minor backroom player, made no report. To him, it was a very bad day. He had already been briefed about the problem in a private meeting with Jean Olivier and was by now trying to think of solutions. "I had the advantage of having had the body-punch earlier." His first hope was that the telescope could still be useful, since the mirror did produce bright cores amid the fuzziness. Moreover, he knew that there was far more demand for telescope time than the time available. Like Bahcall, he recognized that even without sharp images they could still do cutting-edge spectroscopy one hundred percent of the time.[25]

Nor were O'Dell's hopes mere pipe dreams. Back in 1983 Ed Weiler had proposed that NASA stop talking about bringing the telescope back to the ground for servicing and instead plan on doing all maintenance and repair in orbit, using astronauts and the shuttle.

After taking over for Nancy Roman, Weiler had spent most of the 1980s and more than 80 percent of his time wrestling with Hubble issues. He had started his astronomy career working under Lyman Spitzer at

Fig. 5.4. Ed Weiler. (Photo courtesy of Ed Weiler)

Princeton, operating and doing research on the Copernicus space telescope. When Nancy Roman offered him a job at NASA headquarters, however, he made the decision to go into management. Having worked closely with Spitzer, Weiler realized that he could never be that good a scientist. "Whatever I want to do I want to be the best at it," he explained. "I decided it was time to give up on research and go make other people's research possible."

Prior to Weiler's proposal, the agency had been confused about how to maintain Hubble. Very little thought had gone into understanding and comparing the realities of the return-to-Earth versus the servicing-in-orbit scenarios. Moreover, until Weiler had raised the point, no one had recognized how difficult it would be to get Hubble back in space once it was brought back to Earth. As he warned the science working group in February 1984, "If we bring [Hubble] to the ground it'll never fly again."

Weiler not only got his servicing-in-orbit plan approved, he also got funding for NASA to build a clone of WF/PC, just in case anything went wrong with the first instrument, as well as a new spectrograph, dubbed the Space Telescope Imaging Spectrograph (STIS), and an infrared camera, dubbed the Near Infrared Camera and Multi-Object Spectrometer (NICMOS).[26] In subsequent years, the funding for the WF/PC-2, STIS, and NICMOS had risen and fallen, depending on NASA's political circumstances. Yet Weiler had kept all three alive, and by 1990 all three were on schedule for launch and installation on Hubble, the clone in 1993 and STIS and NICMOS in 1996.[27]

Thus, following the initial depressing reports came much more hopeful presentations from the scientists in charge of building these second-generation instruments. John Trauger, the project scientist for WF/PC-2, explained that correcting the aberration would be for him a relatively simple affair. If he could be told the exact prescription, he could essentially adjust WF/PC-2's optics and make it see perfectly. Both Rodger Thompson of NICMOS and Bruce Woodgate of STIS noted that they could do the same for their instruments. All they needed was the precise amount of correction.

Even as Trauger, Thompson, and Woodgate were giving their hopeful reports, Weiler was joining Fisk, Olivier, and project manager Doug Broome for the press conference that would announce the situation to the general public. When it finally began at 1 pm, the science working group stopped what it was doing so that everyone could gather around television sets and watch.

This press conference was reminiscent of a Shakespearian tragedy. Though Olivier, Broome, and Weiler tried very hard to give a clear, accurate, and dispassionate view of the situation, they also could not help but note its catastrophic nature. Worse, though Weiler especially was very clear about the fact that the telescope's spectrographs still worked and that the WF/PC clone could easily repair the aberration and was scheduled for launch in only three years, most of the reporters in the room were only interested in finding out how NASA had failed again and how Hubble had essentially become a huge waste of money. To too many in the media, the only thing that mattered was Weiler's assessment of the main camera: "We feel right now that there's probably no real science that we can do with the Wide Field Camera at this time." Hubble's main camera was out of focus.[28]

The newspaper and media reports over the next few weeks reflected this pessimistic conclusion. In front page story after front page story Weiler's comment was repeated, with headlines that were as depressing: "Defect Ruins Focus of Space Telescope," "Hubble Telescope Crippled," "Faulty Mirrors Make Hubble's Views Fuzzy," "Hubble Hobbled: Stars A Blur." The *Los Angeles Times* was blunt in its disgust. "The world now sees the inglorious result of NASA's laxity and ineptitude." Or as Maryland senator Barbara Mikulski pointedly complained, "Are we going to keep ending up with techno-turkeys?"[29]

It was hardly surprising that the press and public were so quick to condemn NASA and its officials. Only four years earlier the space shuttle *Challenger* had exploded at launch, and during the investigation of that tragedy the public had learned that NASA's management had refused to listen to the warnings of its engineers. Moreover, the cost overruns and dishonest budgeting by NASA in building Hubble had not been forgotten. Past NASA officials had misled Congress and the public about what it would take to build Hubble. Why should anyone believe NASA officials now, especially since Hubble's mirror error appeared to be such a basic mistake?

What made it even more difficult for Weiler and his NASA compatriots to sell their case was that they had nothing to show to alleviate the press's skepticism. Though the mirror was out of shape, the early images from WF/PC did suggest that Hubble was still able to take pictures as good as if not better than anything available on the ground. Yet, the long and still incomplete alignment process combined with the reluctance of the scientists to allow any public relations imagery had prevented the project from imaging anything of interest. Thus, as Sandy Faber noted as she watched the press conference, "They keep describing pictures with their hands but a picture would be worth a 1,000 words."

NASA's credibility problem was made even worse when NASA's independent panel to find out the cause of the spherical aberration released its findings. Chaired by Lew Allen, director of the Jet Propulsion Laboratory in Pasadena, the panel was formed immediately after the June 27 press conference.

When the Allen Commission arrived on July 25, 1990, at the Danbury plant where the primary mirror had been ground, they found the reflective null corrector exactly as it had been almost a decade earlier. Amazingly, in the nine years since Perkin-Elmer had finished polishing Hubble's primary mirror, the company had left the ceiling tower room where the reflective null corrector was installed untouched and sealed. Although the company had been sold to Hughes Aircraft in the interim, the new management continued to keep it locked and protected. Up on the ceiling and out of the way, there had been no reason to disturb it.

Very quickly the investigators discovered the three extra washers still holding the bottom field lens in place. Very quickly they realized that the only reason the technicians had added those washers was that the laser hitting the top of the measuring rod had provided them incorrect information. A careful inspection revealed that the nonreflective cap on the top of the measuring rod was not so nonreflective. In punching out its center hole, the technicians had mistakenly chipped off some of its nonreflective paint. Thus, when the laser beam had been eased into position, its light hadn't bounced off the end of the rod through the center hole in the cap but off the cap itself. Since the cap sat 1.3 millimeters higher than the end of the rod, the laser had given the technicians an inaccurate distance, causing them to position the bottom field lens incorrectly.

That 1.3-millimeter difference in distance meant that when Bud Rigby's polishing team used the null corrector to measure the shape of the primary mirror, they were also measuring it incorrectly. Though they had given the mirror a perfectly precise shape, they had done so to the wrong value. What they thought was the proper curve was actually "too much flattened away from the mirror's center."

In its investigation the Allen Commission also discovered Albert Slomba's ignored test results with the small refractive null corrector. The investigation further uncovered evidence that an *additional* test null corrector had been used repeatedly throughout the polishing process to "aid in the alignment of the [reflective null corrector]." Though this additional corrector's fringes had clearly shown that something was wrong, they were ignored. As the report noted, "The fringe pattern was apparently not fully evaluated at any time during the test period."

During its inquiry the panel interviewed all the important individuals at Perkin-Elmer during those early years, including Bud Rigby and Lucien Montagnino. The best explanation any of these engineers could give for why they had ignored the results from the smaller test null correctors was that they were far less precise than the big reflective null corrector. They simply hadn't believed the results.[30]

■ ■ ■

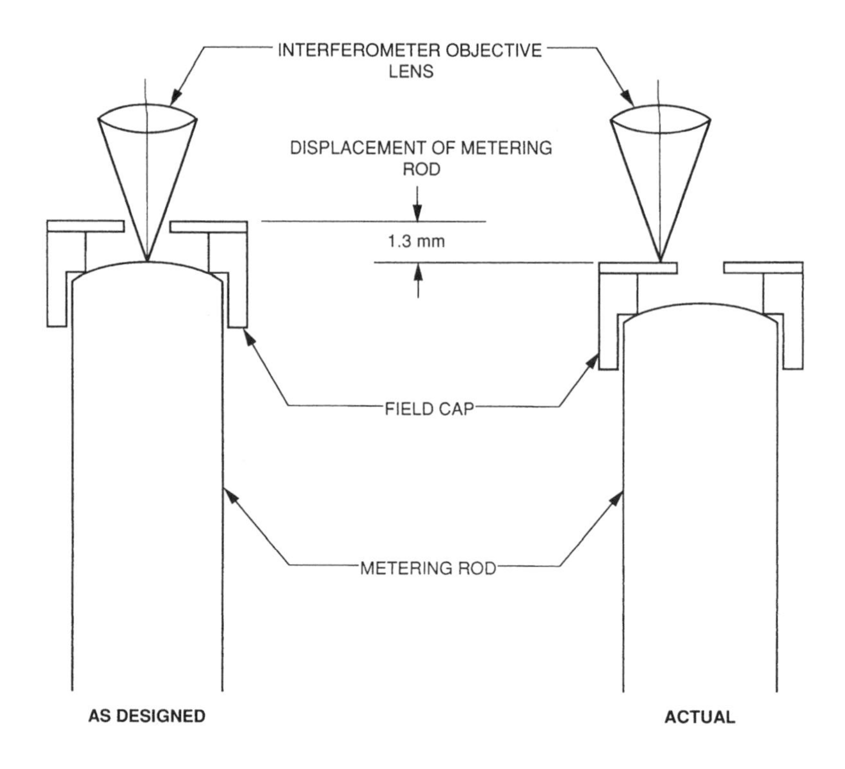

Fig. 5.5. This diagram shows how the 1.3-millimeter error was introduced when the laser reflected off the field cap rather than the end of the metering rod. (Diagram courtesy of NASA)

Amidst this dismal situation there still was hope. Fixing the WF/PC clone and the other second-generation instruments seemed straightforward. All these instruments needed was to have the optics of their mirrors changed to compensate for the error in the primary mirror. As Weiler noted to reporters on June 27, "If we have a positive aberration, we put in a negative aberration and basically cancel it out."

Yet, though NASA's plan would fix the focus problem for all future instruments, including the main camera, it did nothing for the instruments already on board Hubble: the Faint Object Camera, the Faint Object Spectrograph, the High Resolution Spectrograph (now called the Goddard High Resolution Spectrograph), and the High Speed Photometer. Though all four were designed to be removed and replaced by astronauts, opening them up and changing out parts to correct the focus problem was considered impossible. And replacing these instruments entirely by 1993 was too costly. Congress was simply not going to give NASA the funds.

As for the second-generation instruments, both the spectrograph on STIS and the infrared camera on NICMOS were not scheduled for launch until 1996, almost halfway through Hubble's fifteen-year operational life. To the scientists, waiting until then to get these capabilities back meant that the telescope would remain severely crippled for far too long.

One scientist in particular was unhappy about this situation. Holland Ford, a member of the Faint Object Spectrograph team who had been working full time at the institute since 1982, did not like the fact that NASA seemed to be making no effort to fix the other instruments. From his perspective, the institute had a chartered obligation to step forward and put forth its own solution. Moreover, he believed that the most effective way to make something happen in a crisis was to form a "tiger team" made up of the most innovative, informed, and experienced people and let them brainstorm the problem. "In a crisis you want to get the very smartest people you can," he explained.

He started raising the idea at institute staff meetings. Initially Riccardo Giacconi, the institute's director, was on vacation, so no action was taken. At the third meeting, however, Giacconi was back, and immediately agreed, telling Ford to go ahead and form his tiger team. As Giacconi stated later, from his perspective it was "time for the scientists to take their destiny into their own hands."[31]

That Giacconi was unafraid to make demands on NASA was not surprising. Giacconi was known as a man with incredibly high standards of excellence who demanded the best from everyone. He was also known as a man with a forceful personality who liked to get what he wanted. "Riccardo's approach to a meeting is to first roll a hand grenade through the door," noted Sam Keller in 1985. Or as Bob Bless said after Giacconi put on a particularly spectacular performance at a science working group meeting in October 1985, "Riccardo missed his calling. He should have been an actor." In fact, Giacconi had been specifically chosen by the AURA board of directors in order to guarantee that the institute had someone in charge who was willing and able to defend it while also making sure that nothing prevented it from operating at the highest level possible. As Giacconi himself put it, "[They believed that] some of that lack of suavity which I had always had would actually be useful to fight NASA."[32]

This "lack of suavity" became immediately useful when Giacconi took over as director in 1981. Even before the institute was established NASA had awarded the telescope's software operations contract to TRW, and had expected the institute to accept what was given them, no questions asked. Instead, Giacconi and his team reviewed the software's usability, found it inadequate, and demanded and were successful in getting it extensively rewritten.[33]

In fact, it was Giacconi's arrival that helped speed O'Dell's departure in the early '80s. Giacconi's strong presence made O'Dell's Marshall-based job as project scientist superfluous. And though O'Dell's replacement, Bob Brown, was successful during his two-year reign in making sure the telescope was capable of tracking and imaging planets, the existence of the institute and Giacconi also served to handicap his ability to control the project.

In truth, once the institute came into being its director could not avoid automatically becoming the telescope's de facto project scientist. He had both the resources of an entire facility to back him and the responsibility for operating the telescope once launched. And Giacconi could not help but accept that responsibility fully. "If [Hubble] fails, we fail," Giacconi emphasized in 1984.[34]

So, when Ford proposed that the institute put together its own panel for solving Hubble's focus problems, Giacconi was quite willing and able to use the institute's resources to do so. Even as Ford organized the Hubble Space Telescope Strategy Panel (as it was now named), Giacconi began his own campaign to persuade NASA to consider repairing all of Hubble's instruments as soon as possible.

Meanwhile, the Strategy Panel, co-chaired by Ford and Bob Brown, did get the best and brightest, including scientists like Lyman Spitzer, Sandy Faber, Chris Burrows, Ed Groth, and Roger Angel. Angel had been a member of the Allen Commission and was one of the world's experts on building telescopes.

Their plan was to consider any idea or proposal, no matter how crazy. They also followed the advice of Bob Brown. "Holland," he told Ford,

"We have to move this extremely fast." If they took too long coming to a decision, they would lose momentum, as other crises shifted people's attention away from them. Beginning in August 1990, they met four times in the space of two months, hashing out a range of possible solutions, some practical and some absurd. They considered attaching either a large corrective mirror or a large corrective lens to the front of Hubble. They weighed the idea of deforming the primary mirror enough to correct the aberration. They thought about replacing the telescope's secondary mirror. They even discussed the idea of having astronauts recoat the primary mirror, adding enough extra material to its outer edges to compensate for the overpolishing that had made it too flat.

Unfortunately, by the panel's second meeting on September 3 and 4 in Garching, Germany, most of these solutions seemed unworkable. Throughout the first day they listened to a variety of depressing reports. Roger Angel described the commission's findings, including the improvised addition of washers in the null corrector, which seemed to account for all the error. Then Chris Burrows reported on a NASA committee that was working in parallel to the Strategy Panel. Unfortunately, this committee had concluded that nothing on board Hubble, including its actuators, could correct the aberration for the other instruments. This was followed by a report from Pierre Bely about the jitter produced by the solar panels, noting that even if WF/PC was able to take sharp images, the jitter made it impossible to lock accurately on any object in about one third of the sky.

The bad news seemed never to end. Roger Angel gave a second report, describing how adding a tilted correcting mirror to Hubble was impractical. Such a mirror would have to be five times bigger than could be launched, and even if they could get it into space, the extra weight would cause Hubble to vibrate so much that picture-taking would be impossible. Burrows followed again to explain that inserting an additional thin mosaic of lenses above Hubble's secondary mirror was also too challenging. Matching the thickness of the multiple sections plus getting them placed inside Hubble seemed impractical. Then Murk Bottema of Ball Aerospace noted that though mirrors placed in front of the specific instruments themselves could solve the problem, there didn't seem to be any way of getting them installed. These reports were then topped off with a depressing presentation by astronaut Bruce McCandless, explaining the limits of what astronauts could do in space.[35]

That night, one member of the Strategy Panel went back to his hotel room to mull over the situation. Though Jim Crocker knew they would probably be able to fix Hubble's camera, it was essential to find a way to fix the other instruments also, if only to prove to Congress, the public, and the scientific community that space astronomy was worth it. Support for later NASA astronomy projects almost certainly hinged on making the repair of Hubble one hundred percent successful.

An engineer by training, Crocker's dream since childhood had been to build rockets. Though his first job out of college in the early 1970s had been working on Skylab, the late '70s saw a decline in space research, putting him out of the business. Only when he was hired by the Space Telescope Science Institute shortly after its formation in 1981 was he able to get back to doing space work. His job at the institute had been to provide the scientists detailed technical information about the science instruments so they could maximize their observations.

As he worried about the problem, he went to take a shower. The European shower plumbing was unlike the typical American arrangement. Instead of a fixed showerhead high in the wall above the bathtub faucet, the showerhead was fixed to a vertical pipe and drew its water directly from the bathtub faucet through a flexible tube. The showerhead itself could be positioned by sliding it up and down along the vertical pipe. The head also had a pivot arm so that it could be tilted outward at any angle as well as turned either left or right.

During the day, the maid had slid the showerhead down to the bottom of the pipe, and folded it up against the wall. As Crocker got into the shower he slid the head up and pivoted it out. And then he looked at it in wonder.

He thought: what was their actual problem? As Murk Bottema has said that day, though they knew how to correct the aberration with mirrors, they hadn't been able to figure out a way to get those mirrors into place in front of the instruments. Crocker pivoted the showerhead up and down and left and right again, and suddenly had a vision of several different "showerheads" swinging into position and carefully placing their corrective mirrors into the light path between Hubble's secondary mirror and the science instruments.

As he stood there holding the showerhead the idea strengthened. What if they sacrificed one first-generation instrument, say the High Speed Photometer, and inserted this replacement corrective mirror unit into that empty bay? In that case the astronaut's work would be simple and straightforward. Instead of having to climb into Hubble's barrel and do very complicated work wearing thick clumsy gloves, the astronaut would simply pull out the old unit and slip in the new unit. It was exactly the kind of maintenance on Hubble that they "had planned for decades."

Inside this unit Bottema's corrective mirrors would sit at the ends of these movable arms. Once installed on Hubble the arms would automatically unfold—like the showerhead in his hand—and place the corrective mirrors into the light paths for the Faint Object Camera and the two spectrographs.

The next day Crocker presented his proposal to the Strategy Panel. Because he was suggesting the "heresy" of sacrificing one instrument, he wasn't sure how the scientists would respond. Thus, he focused his pitch toward Bruce McCandless, whom he believed would instantly understand the practicality of it.

Crocker was right. "Bruce said, 'Oh, yeah, that'll work.'" But then, so did everyone else. The discussion quickly shifted from trying to think of a way to fix the problem to figuring out how to implement Crocker's idea. For example, Bob Brown quickly noted that they already had a dummy instrument unit, called the Space Telescope Axial Replacement, or STAR, built in the '80s to fill any one of Hubble's instrument bays should a scientific instrument fall so far behind schedule that the space agency was forced to launch the telescope without it. Maybe this would provide them a cheap, quick, and easy way to build this new unit.[36]

The result: the COSTAR, or Corrective Optics STAR. In the end they did not use STAR, since it was cheaper and faster to build a new unit entirely. Moreover, the deployment arms did not pivot out as on Crocker's showerhead, but rose up on a tower and then swung out horizontally. Nonetheless, by the time Riccardo Giacconi presented the Strategy Panel's final report to Charles Pellerin, director of NASA's Astrophysics Division, at the end of February 1991, the entire engineering and scientific community was behind it.

Pellerin's relations with the scientific community, like Nancy Roman's, had not always been smooth. Like her, he had a tight budget, and was often forced to say no, even when he was in favor of a project. For example, when Bob Brown was Hubble project scientist Brown discovered that the software to operate Hubble did not have the capability for successfully tracking planets. To get it rewritten, however, was going to cost money, about $4 million. When Brown presented his case to Pellerin, he had not, as Ed Weiler noted, "greased the sled," let people know in advance his needs and broken those needs down by priorities, so that Pellerin would have some wiggle room in trying to find him the extra money. Worse, Brown showed up late in the budget cycle, after Pellerin's budget had been set for that fiscal year. As Weiler remembered, "Charlie . . . came out of that meeting incensed. They were really upset. They were almost yelling." It required careful negotiations over the next few months by Weiler and others to convince Pellerin to find the cash for the necessary work.[37]

In the case of Hubble's repair, however, Pellerin was very sympathetic. He wanted the telescope fixed, not only because he considered it important to save NASA's reputation but because he thought it important for the country and for science. He had had no trouble finding $50,000 to fund the Strategy Panel's work. To find $24 million to fund its suggested solution, however, was going to be much harder. He and his budget analyst, Greg Davidson, went over the $750 million astronomy budget and shifted money from a variety of other projects, with most of the cash coming from the early construction budget for what was eventually to become the Chandra X-ray Observatory.[38]

As elegant an idea as COSTAR was and as eager as everyone was to build it, construction was not simple. COSTAR had to redirect light from Hubble's secondary mirror to three different instruments, without blocking the path to each or to WF/PC-2. Moreover, two of the instruments needed two pairs of mirrors instead of one. All told, COSTAR carried ten mirrors and four adjustable pickoff arms, with the arms capable of deploying perfectly and then holding their position to as little as one millionth of a meter.[39]

Moreover, during construction Chris Burrows noted that the alignment of the mirrors, on both COSTAR and WF/PC-2, had to be perfect, and that it would be impossible to determine that alignment on the ground. Thus, it was necessary to include the ability to make minute sideways adjustments to three of COSTAR's pickoff arms, once they were deployed and in position, as well as in the corrective mirrors inside WF/PC-2.

Nor was COSTAR the only fix necessary to bring Hubble up to specifications. The solar panels, built by the European Space Agency as their share in the project, also needed to be replaced in order to stop the telescope's jitter. In addition, in the three years leading up to the repair mission three of Hubble's six gyros had failed. There was also the crash of two of the six memory units in the telescope's computer, as well as the failure of two magnetometers that were used by the telescope to find its orientation in space. All had to be replaced or repaired.[40]

Thus, on December 2, 1993, began what was to be one of the most demanding shuttle missions ever. *Endeavour*'s crew of seven had twelve days to catch Hubble, dock it in the shuttle's cargo bay, and then complete five day-long spacewalks of incredible complexity. No single previous spaceflight had attempted so many spacewalks.

Nor was there any margin for error. When NASA officials had been mulling over the repair mission they had considered splitting it into two missions because they didn't think there would be enough time to do everything. In the end, lack of money made a second flight impossible. Instead, they decided to try for it all in one mission, knowing that if any specific task took even slightly too long they would not have the time to do everything.[41]

After two days of orbital maneuvers to rendezvous with Hubble, astronaut Claude Nicollier used the shuttle's robot arm to capture the thirteen-ton telescope and carefully berth it in *Endeavour*'s cargo bay.

Then began the marathon of spacewalks. On the first, Story Musgrave and Jeffrey Hoffman spent almost eight hours replacing three failed gyroscopes, two electronic control units, and eight fuses. Everything went like clockwork, except when Hoffman tried to close the doors to the aft bay where the gyroscopes were installed. For some reason the latches would not line up. After struggling unsuccessfully to get them to latch, the two men left them partly closed and went on to finish all their other scheduled work. Then they returned to the doors, and by carefully pushing, pulling, and fiddling they were finally able to get the latches to catch.

The next day Kathryn Thornton and Tom Akers removed and replaced Hubble's two solar panels. Once again, everything got done, but with several annoying glitches. Thornton's radio failed to receive communications from the ground, and could only hear Akers's voice. For the entire spacewalk she needed him to relay her instructions from mission control. Then, the first panel they were to remove was so warped that it would not roll up on command from the ground, making it impossible to store it in the cargo bay and return it to Earth. Instead, the two astronauts detached it unfurled; then Thornton held the 352-pound panel until sunrise. With the extra light she carefully tossed it overboard over the Sahara Desert, watching it flap away like a giant bird.

On the third day Musgrave and Hoffman finally replaced WF/PC-1 with WF/PC-2. To everyone's relief, pulling out the old unit and installing its replacement went smoothly. When the astronauts used the robot arm to take them to the top of Hubble to replace the two failing magnetometers, however, they discovered that the cover on one old magnetometer was coming off, leaving a potential for contamination, since it was located so close to the telescope's open front end. Since the magnetometers had not been designed for orbital replacement, the plan had been to install the new magnetometers and leave the old ones untouched. The loose cover required some additional unplanned repairs, for which the astronauts would do some improvisation over the next few days.

The fourth spacewalk had Thornton and Akers remove the High Speed Photometer and replace it with COSTAR. Just as the gyroscope doors had caused Musgrave and Hoffman problems two days earlier, the doors to COSTAR's bay were reluctant to close, requiring the two astronauts to struggle with them for a time. They then replaced the telescope's memory units. They also grabbed some insulation off of a handhold on WF/PC-1, now safely stored in the shuttle cargo bay for return to Earth. If all went well, this insulation would be used to seal the uncovered old magnetometer on the last spacewalk the next day.

Finally, on December 9, Hoffman and Musgrave completed the installation of the replacement solar panels by attaching eight new connector cables to each panel's drive units. The new solar panels were then unrolled. Once again, not everything went as planned. First, as they unscrewed a unit, a screw got loose and began floating away. Nicollier quickly used the robot arm to maneuver Musgrave out so that he could snatch the screw before it got away. Then, one of the new solar panels wouldn't swing out from its stowed position when ordered to do so by

Fig. 5.6. The last spacewalk during the 1993 servicing mission to Hubble. Story Musgrave, attached to the robot arm, is about to be elevated to the top of Hubble to install the makeshift cover over the old magnetometer. Jeff Hoffman can be seen in the cargo bay. (Photo courtesy of NASA)

ground controllers. The astronauts had to manually position it perpendicular to Hubble.

Yet, the two men still had the time to ride the robot arm to the top of Hubble and install their makeshift cover to the old magnetometer, fastening it in place with the same ordinary plastic ties used by car mechanics and computer technicians to hold wire bundles together.

On December 10, Claude Nicollier gently lifted Hubble out of *Endeavour*'s cargo bay. At the Goddard control center controllers then began the slow process of opening the telescope's aperture door. Once this was accomplished, they gave the go-ahead for the telescope's release. Nicollier had the robot arm let go, and Commander Richard Covey gave *Endeavour*'s guidance thrusters two short bursts to ease the shuttle away at about one foot per second.

Hubble was once again on its own in orbit. In one of the most spectacular space missions ever, the crew of *Endeavour* had successfully demonstrated that the on-orbit maintenance of an unmanned spacecraft by humans—as first conceived by Hubble's creators in the late 1960s—was not only possible, it was very doable.

The question remained, however, whether Hubble was fixed.

6

"New Phenomena Not Yet Imagined"

Four days after the shuttle returned to Earth, it was once again time to open Hubble's shutter. This time, unlike "first light" in 1990, no one from the press was invited to Goddard to witness it. In fact, this time there was no fake public relations event at Goddard at all. Instead, everyone who could get permission or was important enough not to need it was packed into the same control room at the Space Telescope Science Institute where Chris Burrows and Roger Lynds and a small handful of people had been three years earlier. And rather than a host of television cameras recording everything that happened, a single cameraman had been brought in, merely to record the event for posterity.

"Second light" was also different in that the people operating Hubble had had three years to work out the telescope's kinks, and could quickly get the new instruments and solar panels up to speed and working. For this reason, it took only days to get the telescope prepped, not four weeks.

On December 18, 1993, WF/PC-2 opened its shutter, and took several exposures of a single 12th magnitude star in the northern constellation Camelopardalis. Once again, Chris Burrows was there, standing on the edge of the packed crowd of people hovering over the computer, awaiting the appearance of the first image. Next to Burrows was a very nervous Ed Weiler, whose decision to advocate on-orbit maintenance and the construction of second-generation instruments had made the shuttle repair mission possible. In fact, it was exactly ten years to the day since he had written his white paper on how Hubble should be maintained.

Near the front and center of the crowd was Jeff Hester, a scruffy and tall thirty-five-year-old man, with long hair and beard, who looked more like he should be a mountain man in the Grand Tetons than an astronomer looking at stars. Eight years earlier, Hester had been a young astronomy student about to start a research postdoc at the University of California in Santa Cruz when he had gotten a call from Jim Westphal, asking him if he would be interested in joining the WF/PC team instead. Because Hester had extensive experience programming digital astronomical images, Westphal thought he was particularly suited to work on Hubble.

Hester in turn considered this too great an opportunity to pass up. Hubble was then scheduled for launch in about a year (the fall of 1986). He could help get Hubble launched and operating, and then be in the front line to use it for a year before completing his two-year postdoc and moving on.

Instead, following the *Challenger* accident, Hester as well as everyone else on the project spent the next five years working to get Hubble off the ground. It wasn't until 1990, just before Hubble's launch, that Hester finally took a professor's job at Arizona State University. With the telescope about to reach orbit, he finally expected to spend the next few years reaping the benefits from those years of hard work.

Once again things did not work out as expected. When the spherical aberration was discovered he along with all the other scientists working on the project worried that it might all have been a waste, and that his career might be seriously damaged, if not destroyed, because of Hubble's fuzzy vision.

By the late summer of 1990 things began looking up, however. Tod Lauer was successful in using computer software to clean up WF/PC's aberration enough to produce some decent images of both Saturn and the Tarantula Nebula, a star-forming region in the Large Magellanic Cloud. The Tarantula Nebula is so large that even at a distance of more than 160,000 light-years it can be seen with the naked eye. Despite the spherical aberration, Hubble was able to resolve hundreds of stars never before seen. When Lauer added his deconvolution techniques to it, the image became even clearer. In fact, despite the aberration in Hubble's mirror, this processed image was sharper than anything previously taken from the ground.

Encouraged by these results, Hester and a team including Jim Westphal, Bob Light, Ed Groth, Jon Holtzman, and Lauer immediately used Hubble to take a picture of Eta Carinae. As they wrote in their August

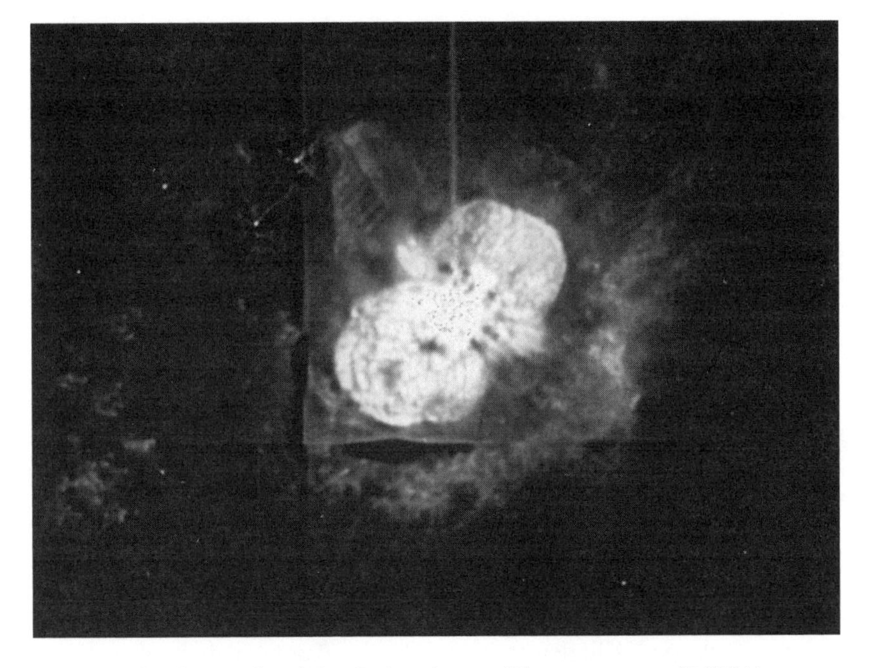

Fig. 6.1. The deconvoluted Eta Carinae image. (Photo courtesy of NASA)

1991 paper, "The opportunity to study a supermassive star at close range during a period when it is highly unstable makes Eta Carinae a unique and important object."

The picture was undeniably ugly and splotchy, with lots of false residual effects left over from the computer processing. Nonetheless, it revealed more detail of the star's Homunculus than ever before. Rather than a "little man," the surrounding nebula formed what appeared to be two bulbous lobes with some jets shooting outward at right angles from the pinched point where the two lobes intersected.

Of all the structures in the image, the one that was most intriguing was the "ladder," pointing northeastward from the center of the nebula. Because the astronomers saw evidence of it in their raw data—before their computer software massaged the aberration out—they believed it was real. What caused it, however, was a mystery to them. "The nature of the ladder is a puzzle," they wrote. The best theory they could come up with was that the "rungs" were ripples inside the jet as it crashed and flowed through and against the cloud of material surrounding Eta.

Fig. 6.2. A digital capture from the video taken during the December 18, 1993, second light, at the moment when the first image began appearing on the screen. Standing from left to right: John Trauger, Jeff Hester, Stefano Casertano, David Crisp, Ed Weiler, Chris Burrows. Sitting in front, left to right: Glenn Schneider, Carl Whittenburg. (Photo courtesy of STScI)

The image suggested strongly that the Homunculus's lobes were formed by a bipolar flow pushing outward from the star's poles. However, that theory did not explain the ladder or jets, nor was the image clear enough to clearly determine where those poles actually were. Furthermore, this model did not do a good job of explaining the structure of the outer portions of the nebula. A second model, which said the entire Homunculus was an oblate shell expanding outward from the star, fit the velocity data of the cloud better than the bipolar jet model.

In other words, despite its success at providing them some new information, Hubble's image of Eta Carinae was still too ugly. As the scientists themselves admitted in their paper, "Much of the structure apparent in the inner 2 [arc-seconds] is questionable." For them to see the star as it really was, Hubble had to be fixed.[1]

With this background, Hester and the others packed themselves into the institute control room in the early morning hours of December 18 as they waited for the "second light" image to process. As it began to appear on the monitor, taking several seconds for the scan to complete, everyone leaned forward, eyes locked on the screen. For Weiler, those few seconds were "the longest three seconds" he'd ever experienced.

Then the image was complete, and everyone in the room burst out into a spontaneous cheer.

There, among a scattering of faint stars, was one bright star, sharp and clear, with almost all of its light confined to a single pixel. As John Trauger gasped, "One bright pixel!" Instead of a fuzzy cloud of haze surrounding a dim core, most of the light was focused to a single point, just as they had always intended. After playing with the contrast for several minutes as well as comparing the different exposures to confirm their initial reaction, the room broke out into more cheers and applause. Someone brought a bottle of champagne, and they all celebrated. "My God, the nightmare of 'Hubble Trouble' is over," Weiler remembered thinking.

Chris Burrows was more skeptical and did not remain for the toast. Instead, as he had done in 1990, he went back to his office and began playing with the data to see if the image was really as good as it had appeared at first glance.

In this case, however, the image was that good. Hubble had been fixed.[2]

■ ■ ■

Everyone at that "second light" event was aware that the semiannual meeting of the American Astronomical Society (AAS) was scheduled to take place in Washington, DC, in only three weeks. They also realized that though it would be great if they could present some results, no one really expected that. There wasn't enough time. Their plan was the same as in 1990: spend several months calibrating WF/PC-2 and COSTAR and release several increasingly good images along the way. The best they could hope for at the AAS conference was for someone from either NASA or the institute to give an engineering update, showing some of these early test images to demonstrate that the telescope had been fixed.

Suddenly, however, things changed. The "second light" image had been taken on December 18. Burrows analyzed it and suggested alignment changes. The engineers at Goddard made these changes and immediately took another test image. Burrows analyzed this and made more suggestions. Another image, more suggestions. Within days they were able to get the telescope aligned, and by Christmas they were able to take a fairly nice picture of the globular cluster 30 Doradus, located inside the Tarantula Nebula. The picture was so good that the scientists went to Ed Weiler and said, let's start doing science. Weiler immediately agreed.

Everyone took a few days off for Christmas while they let the new instruments cool down to the ambient temperature of space, then returned with a rush to take pictures of a host of astronomical objects, including the beautiful spiral galaxy M100, and the Orion Nebula. (See

Fig. 6.3. Supernova 1987a, taken by Hubble's Faint Object Camera, using COSTAR's corrective optics. The debris from the supernova is in the center. The circumstellar ring was produced 30,000 years earlier, when the aging central star went through a violent phase of mass loss, somewhat similar to the eruptions of Eta Carinae seen today. (Photo courtesy of NASA/ESA)

color plates 9, 10, 11, and 12.) "There was great excitement," remembered Jeff Hester. For example, the picture of M100 was taken New Year's Eve. Though everyone was invited to a party at the home of one of the scientists, no one really wanted to be there. "We rang in the New Year," Hester said. Then, "Everybody sort of looked at their watches and said 'All right, it's time to go home now.'" The party ended, but no one went home. Instead, "We all went back to Hopkins and stayed up the rest of the night actually working on the images."

The pace continued to accelerate. Though most of the best images were going to come from WF/PC-2, there were also several pictures from the Faint Object Camera, using COSTAR. Getting good images through COSTAR took more time. Not only did the scientists have to deploy each pickoff arm separately, but they then had to do careful alignment tests for each. Nonetheless, by early January the COSTAR teams had several sharp pictures of their own, taken with the Faint Object Camera, including the 1987 supernova in the Large Magellanic Cloud as well as the 1992 nova in the constellation Cygni.

With such a plethora of good news, the decision was made to put together a presentation at AAS. A special session was quickly arranged. On January 14, 1994, members from the WF/PC-2 and COSTAR teams—including Chris Burrows, Jeff Hester, Jon Holtzman, and Jim Crocker—presented both data and images to a packed room of more than 2,000 astronomers in the main ballroom of the Crystal City Marriott Hotel, just outside of Washington, DC.

Now it was time for the rest of the astronomical community to cheer. That Hubble could clearly show individual stars in the core of M100 as well as the expanding rings surrounding the 1987 supernova illustrated how successful the repair had been. Burrows couldn't help joking, "We

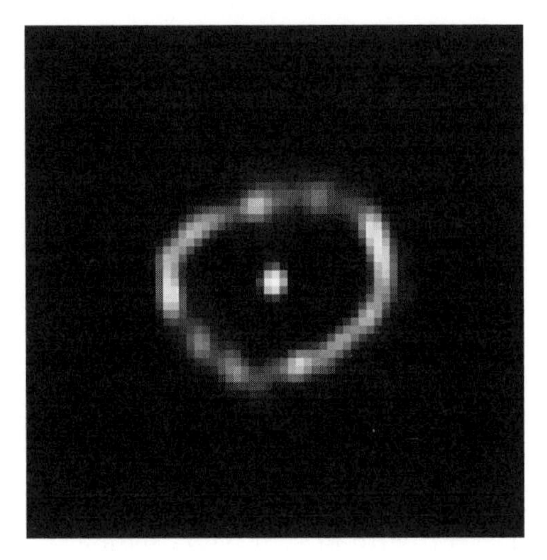

Fig. 6.4. Nova Cygni 1992, taken by Hubble's Faint Object Camera, using COSTAR's corrective optics. (Photo courtesy of NASA/ESA)

got the sign right!" referring to the common mathematical error of mistakenly switching the plus/minus sign in some crucial equation. Then to his chagrin he dropped his microphone, causing more laughter.

Crocker was more direct. "We nailed it—got the prescription just right."

For the scientists, the most thrilling moment might have been when Jon Holtzman showed a graph of the photometric data compiled from 30 Doradus. First he placed a viewgraph on the screen, showing, on its left side, a graph of the typical distribution of stars in a globular cluster in terms of brightness and temperature; the viewgraph's right side was blocked by a piece of paper. Then he removed the paper, showing the same graph for 30 Doradus, as measured by Hubble. It was an exact match. For the scientists this was breathtaking. "It was a one-two punch," Hester remembered. "Not only could we produce spectacular images and see all sorts of detailed structure, but we could do quantitative photometry."[3]

Almost fifty years after Lyman Spitzer had first proposed it, a working optical telescope was finally in orbit. "Three years ago," Spitzer told reporters, "I received news of the defect in the mirror on my birthday, making it one of the worst birthday presents I ever received. The good news [today] offsets that sorrow."[4]

■ ■ ■

This spectacular moment of triumph was only the beginning. Hubble had been designed as a long-term general observatory, available for any astronomer in the world to use. Its primary mission had been set to last a minimum of fifteen years, with the potential to go far longer. Moreover, its modular design, linked to periodic human repair missions, meant that its instruments were going to be repeatedly replaced and improved.

The 1993 repair mission was followed by servicing and repair missions in 1997, 1999, and 2002. On the 1997 mission astronauts installed both NICMOS and STIS, replacing respectively the Faint Object Spectrograph and the Goddard High Resolution Spectrograph. STIS not only had a higher spectral resolution, but it could also obtain multiple spectra across its field of view during a single observation, rather than having to make separate observations for each spectrum. NICMOS meanwhile gave Hubble its first ability to observe the heavens in the infrared. The astronauts also kept Hubble working by replacing two of its gyros and one of its three fine guidance sensors as well as installing a new solid state data recorder.

The 1999 servicing mission had originally been scheduled for June 2000, and was to have included the installation of a new, even more sophisticated camera, dubbed the Advanced Camera for Surveys (ACS). However, when three of Hubble's six gyroscopes failed in mid-1999, the mission was split in two so that new gyros could be installed sooner. This decision was fortuitous, with a fourth gyro failing in November 1999, putting Hubble into safe mode and preventing scientific research. The emergency rescue mission was launched December 19, 1999, and lasted eight days, during which astronauts replaced all six of the telescope's gyros plus a second fine guidance sensor. They also installed a new computer, a new voltage/temperature kit for the spacecraft's batteries, a new transmitter, and a new solid state recorder. In addition the astronauts improvised the installation of thermal insulation blankets when they noticed damage to the telescope's outer layers due to the hostile space environment.

The second half of this split servicing mission was finally launched on March 1, 2002. Once again astronauts replaced a host of equipment, including two gyros, the telescope's main power unit, and the solar panels that had been installed in 1993. Other repairs included the installation of a new permanent cooling unit on NICMOS, which brought the instrument back to life after the cryogen in its original cooling unit had sublimated away faster than expected due to a design flaw. Finally, the astronauts removed the Faint Object Camera and replaced it with the new ACS camera, with a field of view more than twice the size of WF/PC-2 with five times the sensitivity.[5]

The repairable and maintainable design of Hubble—conceived and pushed through by men like Spitzer, O'Dell, and Weiler—had made it possible for Hubble to work. Had it been built as a typical unmanned space probe, the error in the mirror would have made it almost unusable, right from the beginning. Moreover, its lifespan would have been seriously shortened without the ability of astronauts to periodically do repairs.

In fact, better than any other space probe, Hubble has demonstrated the value and necessity of combining human and robot operations in space. Robots and unmanned spacecraft can achieve a lot for much less money, but they do so by sacrificing flexibility and capability. Human spaceflight in turn is expensive and difficult, but the addition of human skills to any robot operation makes the robots far more effective and useful. More importantly, humans can *fix things*, something no unmanned probe can do. Future space explorers will ignore this lesson at their peril.

The payoff from Hubble's design did more than benefit the telescope and space exploration. The willingness of Hubble's builders to adopt new technology also helped change the ordinary world around us. What few today realize is that the charged coupled devices or CCDs specifically developed for both the Hubble Space Telescope and the Galileo mission to Jupiter ended up helping to kick-start a revolution in photography. Following these missions the electronics industry began an intense push to develop both video and still cameras using this CCD technology, and by 2003 the leap from film to digital imaging had become so complete that companies like Kodak, which had made film the standard at the start of the twentieth century, had publicly abandoned film for digital photography at the start of the twenty-first century.[6]

Meanwhile, the scientific rewards since 1993 have been astonishing. As of early 2006 more than 6,000 scientific papers had been published based on Hubble data, totaling about 35 percent of the entire scientific output from all NASA research projects and more than three times the output of its nearest competitors, the Voyager and Viking probes to the outer solar system and Mars respectively. This amazing unending stream of data has not only helped revolutionize the field of astronomy, it has changed our very perception of the universe.[7]

For astronomers, Hubble on a very practical level helped transform the way research was done. Before Hubble and the launch of the first space telescopes almost all the big telescopes in the United States were controlled by private observatories, a situation that served to limit access to those telescopes. It was this circumstance that caused support for the space telescope to split along regional lines, with most of the support coming from eastern astronomers, who generally lacked easy access to the big telescopes at Palomar Mountain and Mount Wilson, and most of the opposition coming from western astronomers, who usually controlled those big western telescopes and felt no urgent need for a general observatory in space.

Today the culture of astronomy is significantly different. Though many of the largest ground telescopes are still privately run with limited access, no longer do a few private organizations control access to most of the world's best telescopes. Instead, many of the biggest and most important telescopes, both on the ground and in space, are general observatories following the model set by Hubble. A public institute similar to the Space Telescope Science Institute operates the telescope. Telescope time is available to anyone in the world, with the time allocated competitively by committees of astronomers chosen from the entire astronomical community. The data, though reserved to each observer for a period of time, are then archived and made available for anyone to use.

More importantly, the manner in which most of today's big telescopes are operated has changed. In the past the astronomers working with one of those private telescopes would generally calibrate and operate the instruments on the telescope and then analyze the data themselves. This meant that a user had to be intimately familiar with each instrument in order to get usable data. Moreover, the data from each instrument were often inconsistent, depending on who was doing the calibration, operation, and analysis.

Hubble could not be used in this manner. Not only was the telescope going to be operated remotely in space, but many of its users were going to be unfamiliar with the intricate nuances governing the operation of a space telescope. To solve this problem the Space Telescope Science Institute had to take on a greater responsibility, doing the calibration, operation, and analysis for the astronomer. The result was that the data produced has been more consistent across the board. More importantly, it is now possible for researchers unacquainted with the operation of a space telescope to obtain usable data from it.

This approach established at Hubble was so successful that the organizations operating many of today's biggest telescopes are following it, both in space and on the ground. "There was a tremendous change in the way astronomy was done," explained Riccardo Giacconi. "Every astronomer in the world, every person in the world can now . . . get access to the best data from the best telescopes."[8]

Then there was the data itself. For example, John Bahcall was finally able to do his optical survey of quasars. Before Hubble, astronomers had a limited amount of information about the nature of these strange objects. Though they resembled stars, their light spectrum suggested that they had to be many billions of light-years away, which in turn meant that to be as bright as they were they had to be among the most energetic objects in the universe. Careful ground-based telescope images showed evidence that quasars were found in the centers of galaxies, though the limitation in resolution caused by atmospheric light scattering made the data hard to interpret. Based on what was known, the consensus among most astronomers was that quasars were large supermassive black holes, each in the center of a galaxy. Bahcall's survey of quasars hoped to clarify these conclusions while trying to determine the nature of those galaxies.

Of the twenty quasars that Bahcall and his team imaged, all turned out to be at the center of some form of galaxy, either ellipticals or spirals. That ellipticals dominated slightly was not surprising to the scientists. What was surprising, however, was the number of quasars embedded in merging galaxies, galaxies with very close companions, galaxies with well-developed spiral arms, or galaxies with strange shapes. The data suggested that as the first galaxies formed in the early universe, the intense gravity of these supermassive black holes served as a trap to pull in matter from multiple surrounding galaxies.[9]

In fact, though data from many earlier ground-based studies had suggested it, Hubble data have proved that black holes are astoundingly ubiquitous throughout the universe. For example, Hubble found that almost every spiral galaxy that has a large bulge surrounding its nucleus has a supermassive black hole in its center. Moreover, the size of each galaxy's central bulge is proportional to the size of that central black hole: the bigger the bulge the bigger the black hole. Later Hubble data have also helped unify the theories that were trying to explain the different kinds of energy that radiated out from the centers of many galaxies, including radio, ultra-violet, x-rays, and gamma rays. Scientists now believe that these emissions are caused by the in-fall of matter into the central supermassive black hole.

Everywhere scientists pointed Hubble it revealed a universe that was as beautiful as it was astonishing. For example, Hubble's first sharp image of the 1987 supernova in the Large Magellanic Cloud was only a precursor to more than a decade of follow-up observations. Since its launch and even before the 1993 repair Hubble had carefully tracked the spectacular changes following this violent death of a star. By 1996 the inner ring surrounding the star was beginning to brighten as it was hit by the wave of material flying outward from the explosion at speeds of 2,500 miles per second. Over the next decade Hubble documented these fireworks as a ring of spots, reminiscent of pearls on a necklace, steadily brightened year by year. (See color plate 23.) In the center of that ring Hubble was also able to resolve the remnant debris from the supernova blast, something no other telescope in the world could see, giving astronomers their first detailed and continuing look of the immediate aftermath of a supernova.[10]

Within our own Milky Way galaxy that first image of the Orion Nebula led to a revolution in the study of star-forming regions and the thin gases between the stars. The best and most well-known example of this research was the startling and spectacular image that Hubble took of the Eagle Nebula, showing towering pillars of gas being illuminated and eaten away by the radiation of a nearby hot star. (See color plates 7 and 8.) The Orion image itself was remarkable in that not only did it show the streams of gas flowing through the nebula in better detail than ever seen before, but silhouetted against this backdrop were strange blobs that Bob O'Dell quickly dubbed "proplyds" for proto-planetary disks. These disks were indisputable evidence that the formation of solar systems was a natural part of the birth of stars. (See color plates 9 and 10.)

Later data from Hubble showed the dust disks around other stars in spectacular detail. For example, the star Beta Pictoris was surrounded by a dust disk; Hubble data were further able to resolve this debris into two disks, one inclined about 4 to 5 degrees from the other. The second disk suggested it was formed by the existence of a planet sweeping up debris into its slightly tilted orbit. Other Hubble data reinforced this conclusion by showing how the main disk of the star was slightly warped, as if the presence of other planets was reshaping the disk as they orbited their way through it.

Hubble's sharp resolution and ability to see faint objects helped astronomers by adding remarkable clarity to their data. More importantly, from their perspective Hubble's greatest strength was that it was a superb and unmatched supplement to the many hundreds of other telescopes they used both on Earth and in space to make observations across the entire electromagnetic spectrum.

For example, the enigmatic and ephemeral mystery of gamma ray bursts, which usually last no more than a few seconds and pop up randomly about once a day anywhere in the sky, had long been a mystery. Discovered in the late 1960s by the first orbiting gamma ray detectors, the bursts were so brief that astronomers had not been able to detect them in any wavelengths other than gamma rays, which meant they had no way to measure their distance or energy output. For all scientists knew, the bursts could be occurring inside our solar system or in the most distant galaxies billions of light-years away.

In the late 1990s astronomers were finally successful in catching the fleeting x-ray and optical radiation of several long bursts and were thus able to measure their distance, using a combination of space- and ground-based telescopes, proving that these events occurred many billions of light-years away. Astronomers then aimed Hubble at these objects and learned that the bursts took place in faint star-burst galaxies, places where star formation is going on at a furious pace and far more aggressively than seen in the universe today. Hubble also helped show that the bursts were linked to unusual supernovae (dubbed hypernovae), in which the core of the star collapses almost instantly, producing an explosion of unprecedented strength.[11]

In another example, in the 1990s when astronomers began discovering planets orbiting other stars using ground-based telescopes, Hubble helped to confirm and refine these results by actually observing the transit of numerous extrasolar planets across the face of their central stars. The telescope was aimed at the bulge of the Milky Way galaxy and, in one long seven-day observation, was able to identify sixteen extrasolar planet candidates, of which at least a third were almost certainly Jupiter-sized planets. Many of these planets were strange and alien, orbiting extremely close to their stars, with one having a "year" only ten hours long at a distance of only 740,000 miles from its sun—just three times farther than the Moon is from the Earth. In other observations, Hubble's high resolution (able to measure the subtle drop in light output as a planet blocked the star's light) was so precise that scientists were able to measure the presence of hydrogen, sodium, carbon, and oxygen in the planets' atmospheres. Hubble was even able to show that these extrasolar planets did not have rings or large satellites like the gas giants in our solar system.

Though Hubble's extrasolar planet findings played a relatively small part in this exciting field, its results showed the incredible value of having an optical/ultraviolet telescope in space. Hubble could quickly obtain data that were difficult if not impossible for ground-based telescopes to get. Moreover, its success at detecting elements in the atmospheres of these incredibly distant planets proved that such an advance in research was even possible.

Closer to home, Hubble has given planetary astronomers a spectacular tool to study the solar system. They have measured the seasonal weather changes in the atmospheres of Mars, Jupiter, Saturn, Uranus, and Neptune. They have discovered auroras above the poles of Jupiter and Saturn. They produced the most detailed image of Pluto yet, even if the resolution of that image is still somewhat vague. And when Comet Shoemaker-Levy smashed into Jupiter in July 1994 astronomers used Hubble to take numerous pictures, showing gigantic impact plumes rising above the planet's surface as well as powerful ripples spreading outward at more than a thousand miles an hour. (See color plates 18, 19, 20, 21, and 22.)

In fact, a complete summary of the knowledge gained by astronomers using Hubble would fill this book and can't possibly be summarized here. All told, in its fifteen-plus years of operation there is almost no area of astronomical research that the telescope has not in some way affected or changed.

Of this vast wealth of new knowledge, however, astronomers generally consider Hubble's contributions to cosmology—those concerning the birth and age of the universe—to be its most significant.

When astronomer John Herschel completed his survey of the heavens in the 1840s, he had catalogued thousands of beautiful but hazy objects scattered everywhere across the sky amid the innumerable stars. Different from planetary nebulae, which often have distinct structures surrounding specific stars and are generally clustered along the plane of the Milky Way, these "nongalactic nebulae" generally looked like fuzzy footballs, spirals, or whirlpools. For the next hundred years, astronomers puzzled over their origin. Not only did no one know what they were, no one was even sure if they were local objects inside our galaxy or distant island universes comparable to the Milky Way.

In the 1920s, the American astronomer Edwin Hubble finally had a tool for finding out. Using the 100-inch Hooker telescope at Mount Wilson, California in 1924, he was able to resolve a large number of individual stars in several of the larger nongalactic nebulae, proving that their fuzziness was caused by billions of stars, not dust or gas. More importantly, in the Great Nebula in Andromeda about three dozen of those stars were variable, with twelve being a special type of variable star called Cepheids. Cepheids are unique in that their intrinsic brightness, or absolute magnitude, is directly related to the length of their pulsations. The brighter they are, the slower they pulse.

Based on the pulse rates of the Cepheids in Andromeda, Edwin Hubble could compute their absolute brightness, and from this the distance from Earth, which he estimated to be about a million light-years. Soon he found Cepheids in a number of other nongalactic nebulae, proving that these too were many millions of light-years away. Though later research has significantly refined the estimated distances to these galaxies, Hubble's work provided proof that the universe was filled with innumerable collections of stars as large as if not larger than our home galaxy, the Milky Way.[12]

This wasn't all, however. Even as Edwin Hubble made this discovery, he noticed a strange phenomenon. The farther away a galaxy was, the faster that galaxy seemed to fly away from the Milky Way.

Hubble discovered this phenomenon by noting the "redshift" of the light coming from these and other distant galaxies, as compiled by other astronomers. If a galaxy or star is approaching the Earth, the wavelengths of light it emits will be squeezed together by that motion, making them shorter and thus shifting the color of that light toward the blue end of the spectrum. And if the star is instead moving away from the Earth, that movement will stretch the light waves, so that its color will shift toward the red or longer end of the spectrum. Since scientists know the exact wavelength at which elements produce lines in a star's spectrum, if they see any shift of those lines to either the blue or red, they know the star is moving and in what direction. The amount of shift will also tell them the speed of that movement.

Hubble found that the spectrum of almost every galaxy was shifted to the red. More importantly, the greater the distance, the greater the redshift. In other words, the universe was expanding.

Scientists understood immediately that this expansion was not actually motion. The Milky Way is not the center of the universe, with all other galaxies fleeing from us in terror. Instead, the space in which everything exists is getting stretched.

The best and most oft-used analogy to describe this phenomenon is that of a balloon. Blow the balloon up about halfway. Then take a felt marker and place dots across the balloon's surface, spaced about one inch apart. Now pump more air into the balloon. As you do so the dots will move away from each other so that when the balloon's size is doubled the space between all adjacent dots will double, going from one inch to two, with dots that started out two inches apart now four inches apart and dots four inches apart now separated by eight inches. In other words, the distances increase linearly.

Should you focus your eye on a single dot as the balloon grows in size, it will appear as if it is at the center of an expansion and all other dots are fleeing from it. Repeat this experiment but focus on another dot and *it* will appear to be at the center.

More significantly, note that *none* of the dots are really moving. Instead, the "space" in which they are embedded is expanding, moving them away from each other as it grows. Similarly, the galaxies in space are not really fleeing from each other. Instead, it is space itself, the very fabric of the universe in which they reside, that is expanding. By extrapolating that expansion backward scientists found that the universe appeared to have a cataclysmic beginning, what astronomer Fred Hoyle labeled contemptuously as the Big Bang.[13]

Determining the rate of expansion would help scientists calculate how long ago the Big Bang happened. However, in the decades following Edwin Hubble's initial work astronomers found it difficult to come up

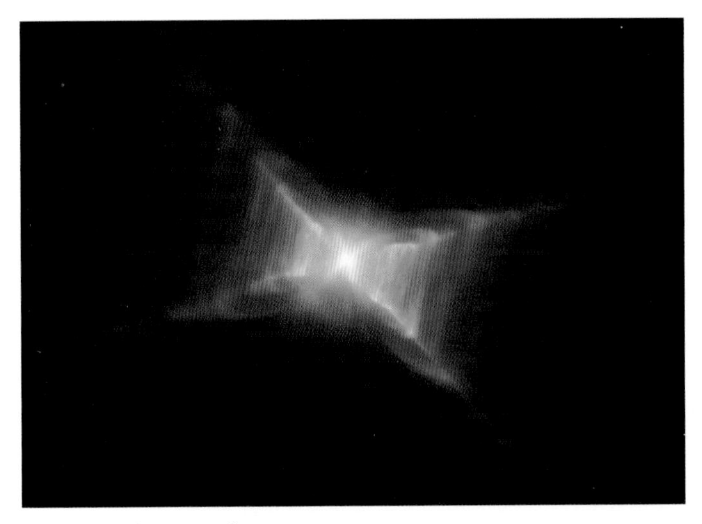

Plate 1: Red Rectangle

Plate 2: Ant Nebula

Plates 1-6 provide a small sampling of only a few of the many spectacular planetary nebula images produced by the Hubble Space Telescope. Each illustrates a different aspect of the complex and still not-entirely understood process that forms these beautiful structures.

The Red Rectangle Nebula exhibits a baffling x-shaped pattern with a spider-web of "ladder rungs" looping from bar to bar, while the Ant Nebula displays two distinct bipolar shells surrounded by a more amorphous outer envelope of strange wisps.

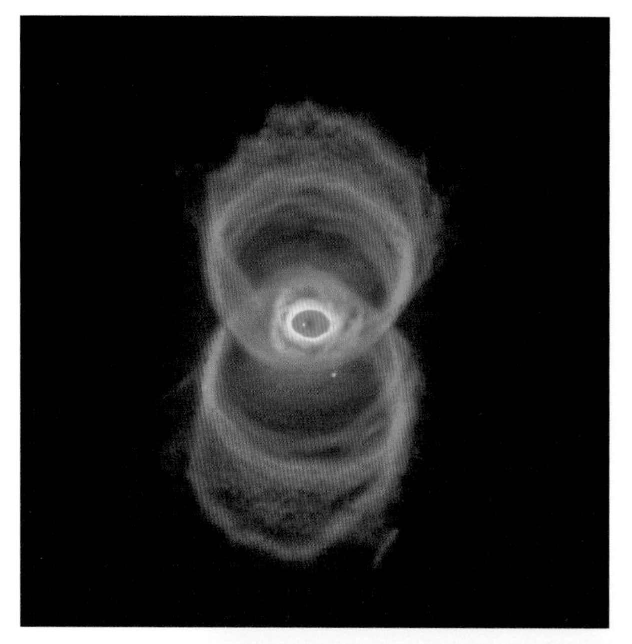

Plate 3:
Hourglass Nebula

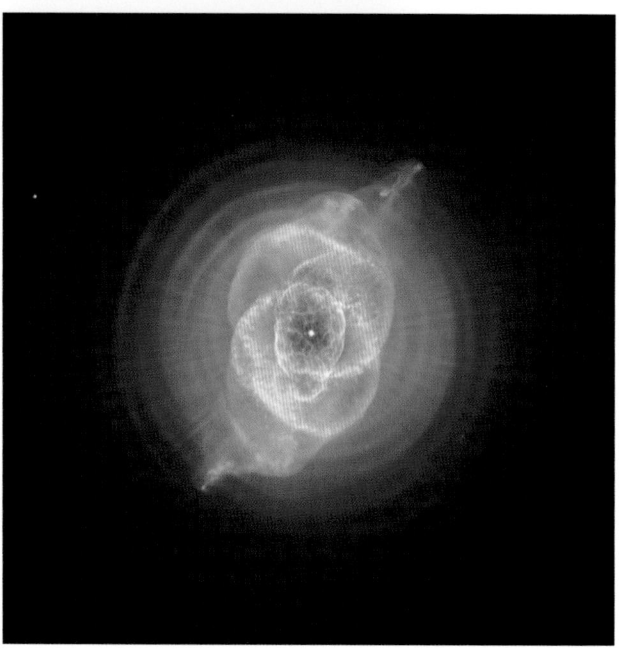

Plate 4:
Cat's Eye Nebula

The Hourglass Nebula shows the classic bipolar shape of two shells expanding outward from the central star, in opposite directions. The Cat's Eye Nebula in turn has an incredibly complex series of at least eleven interweaving concentric shells.

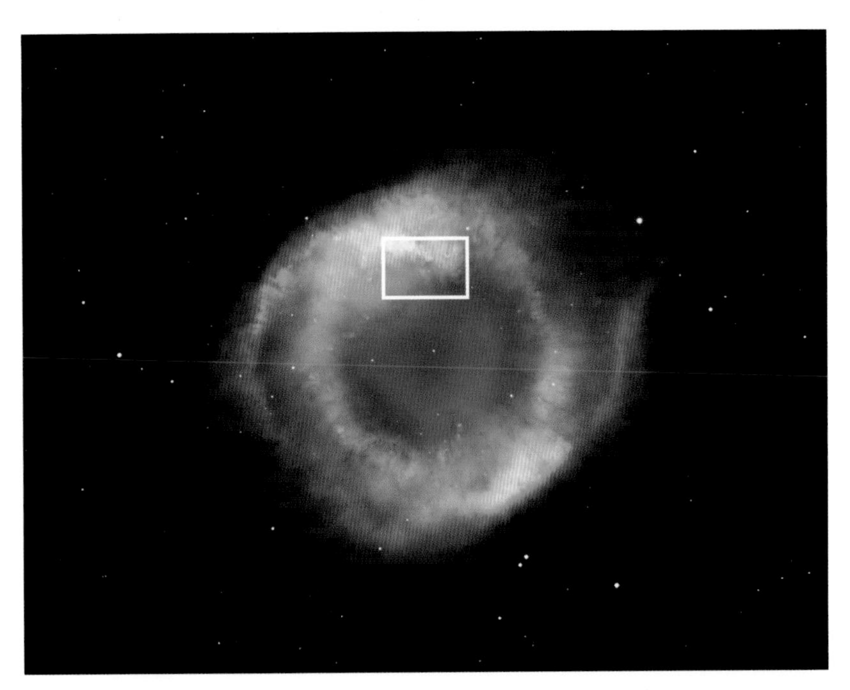

Plate 5: Helix Nebula

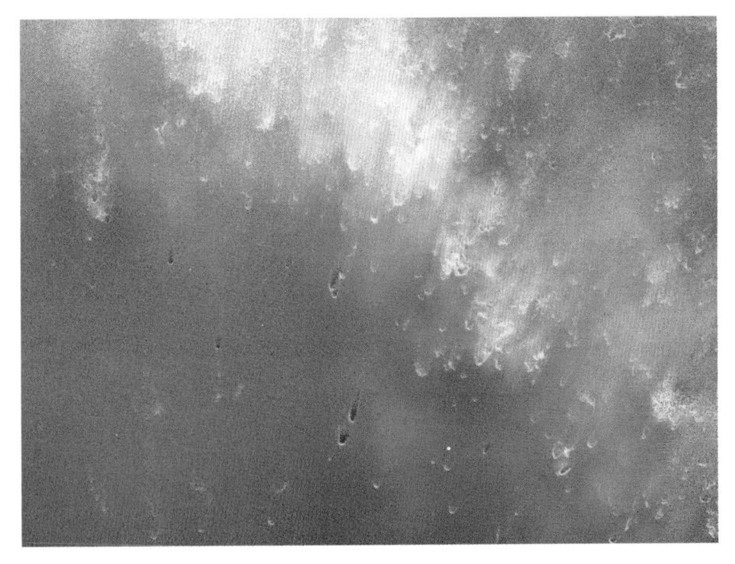

Plate 6: Inset, Helix Nebula

The Helix Nebula appears to be a gigantic shell, with the star's inner radiation eroding the shell away in a process similar to that occurring at the Eagle Nebula (plates 7 and 8) and resulting in the gaseous knots visible in the inset, each about twice the size of our solar system and all pointing inward toward the star, much like comets.

Plate 7: Eagle Nebula

Plate 8: Inset, Eagle Nebula

The Eagle Nebula in the constellation Serpens. When Jeff Hester first showed this image to Ed Weiler, Weiler's immediate reaction was "That'll make the cover of *Time* magazine." Weiler was partly right, as the image ended up on the cover of the international issue of *Time*. The inset shows the top of the leftmost pillar, covering an area about the size of our solar system. Ultraviolet light from several nearby young, hot, and massive stars beyond the top edge of the image is heating the pillar's outer layers of gas. Over time this radiation causes the gas to photoevaporate, eroding away the cloud and thus sculpting out the fingers at the top of the pillar.

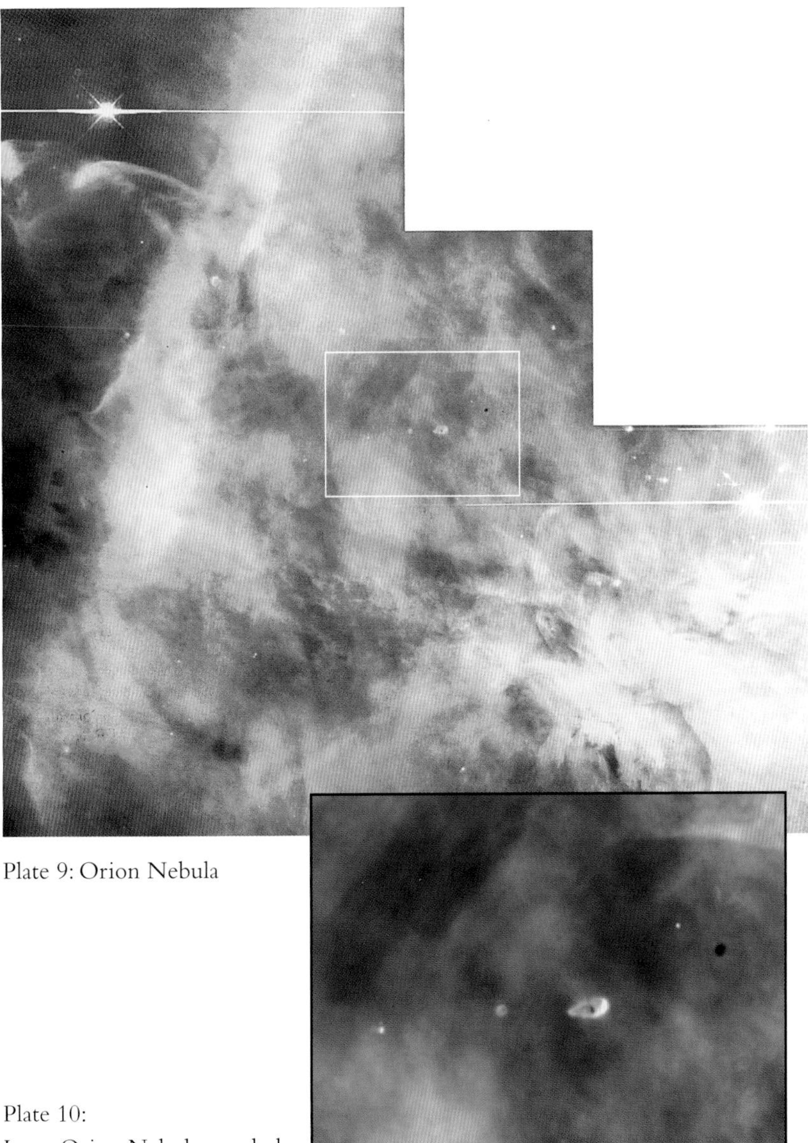

Plate 9: Orion Nebula

Plate 10:
Inset, Orion Nebula proplyds

The Orion Nebula, taken by the Hubble Space Telescope immediately after its repair in December 1993. The inset shows several proplyds. A survey of plate 9 by O'Dell and Zheng Wen of the University of Kentucky showed that of the 110 stars in the image more than half were surrounded by disks. The disklike shape of the surrounding material, rather than being spherical shells, supported the theory that they were protoplanetary solar systems. These data, obtained before the discovery of any extrasolar planets circling ordinary stars, bolstered the theory that planetary formation was common throughout the galaxy.

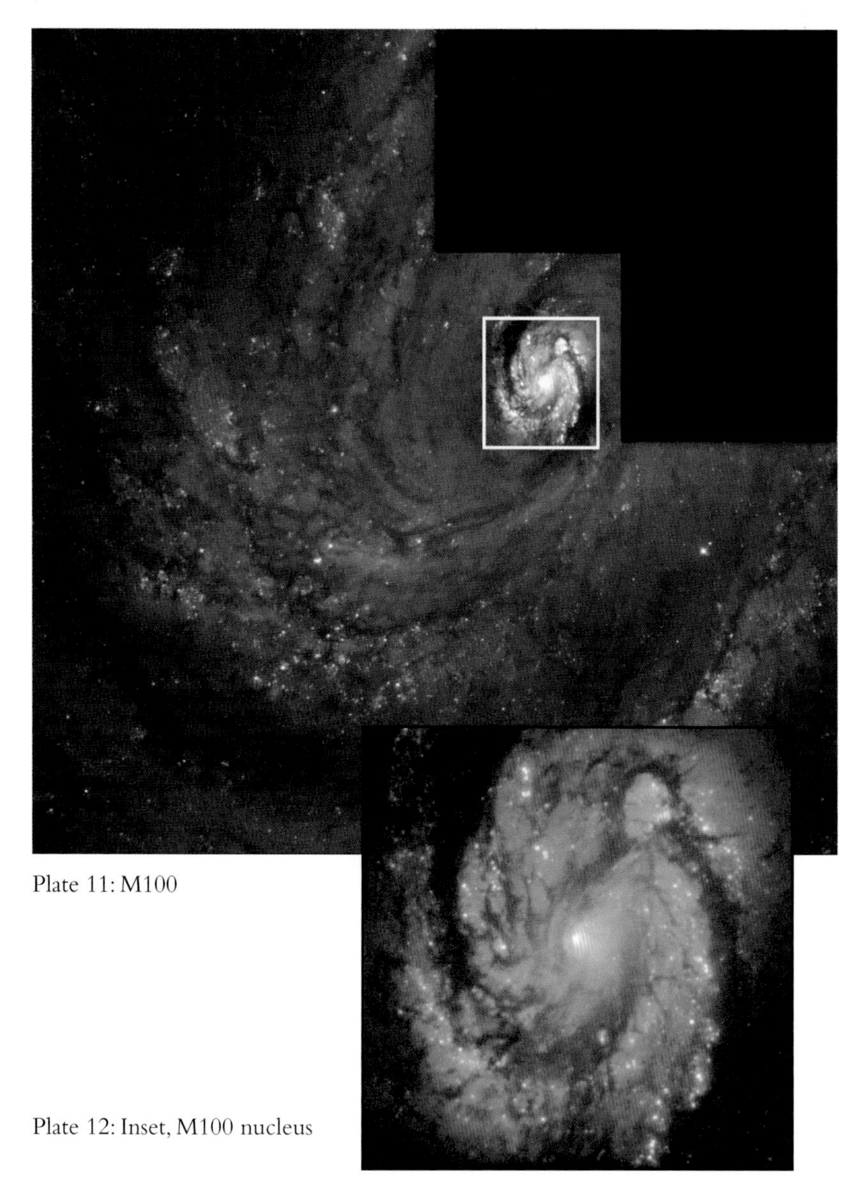

Plate 11: M100

Plate 12: Inset, M100 nucleus

The galaxy M100, located approximately 40 million light-years away in the Virgo galaxy cluster. This image was taken by the Hubble Space Telescope immediately after its repair in December 1993. The inset of this spiral galaxy's nucleus shows remarkable detail, resolving faint structures 30 light-years across. This image's clarity demonstrated that Hubble would be able to identify Cepheid variables in the Virgo galaxy cluster and thereby more precisely measure both the universe's expansion rate and its age.

Plate 13: Sombrero Galaxy

The Sombrero Galaxy, M104, also located in the Virgo galaxy cluster. This mosaic of six images, taken in May/June 2003 with the Advanced Camera for Surveys, not only shows the dust lanes skirting the galaxy's equator, but also resolves nearly 2000 globular clusters surrounding the galaxy in a spherical shell. Closer inspection of the bright nucleus reveals a smaller disk, tilted relative to the larger disk that surrounds it.

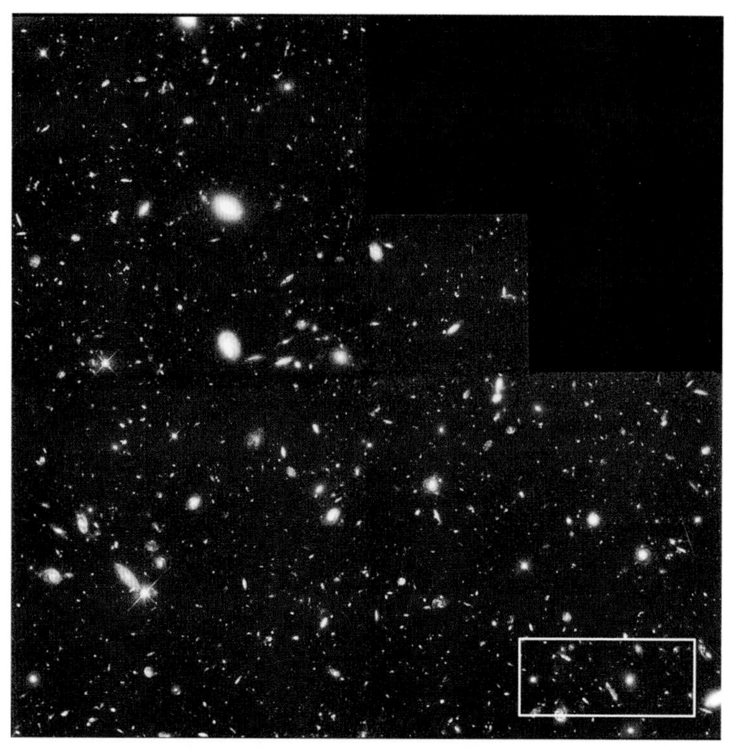

Plate 14: Hubble Deep Field

Plate 15: Inset, Hubble Deep Field

These two deep field images, one taken in 1995 and the other in 2003/2004 (plates 14 and 16), represent the longest and deepest views of the universe yet. Practically everything in both images is a galaxy, not a star. The two insets (plates 15 and 17) show the strange irregular nature of many of these distant galaxies, compared to what is seen in the nearby universe. And though several of the larger galaxies in the insets appear to resemble typical spirals, a closer look shows that the spiral arms are usually less distinct, or even nonexistent.

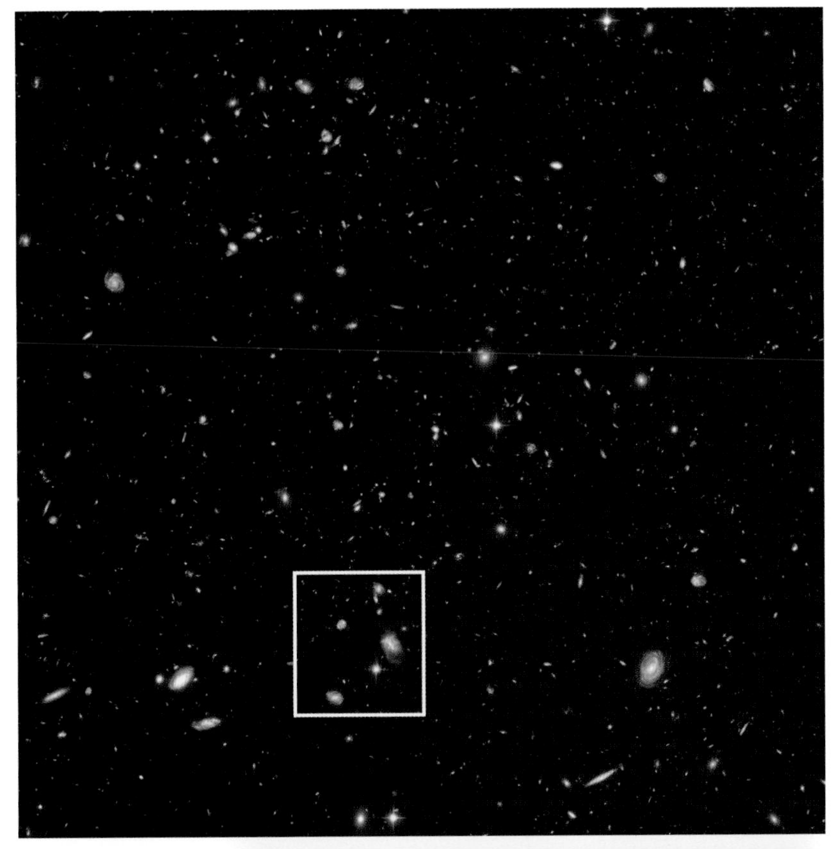

Plate 16: Hubble Ultra Deep Field

Plate 17: Inset, Hubble Ultra Deep Field

Plate 18: Portrait of Martian weather

These planet images, plates 18, 19, 20, 21, and 22, give a taste of Hubble's solar system output.

The Mars image from April 1999 shows a large water-ice storm more than 1000 miles across just to the west and south of the north pole.

Plate 19: Aurora over Saturn's poles

The October 1997 first picture of Saturn's ultraviolet aurora was taken by scientists using Hubble's Space Telescope Imaging Spectrograph as a camera.

Plate 20: Jupiter storms

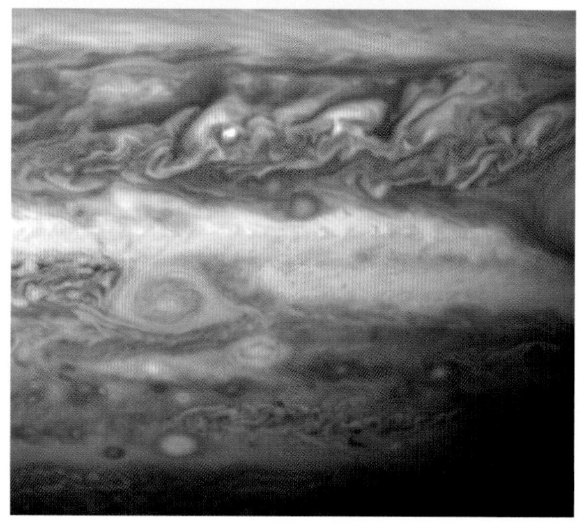

Plate 21: Inset, closeup of Jupiter storms

The inset image of Jupiter's storms, from April 2006, shows both the planet's Giant Red Spot as well as the new Red Spot Jr. to its west. Planetary scientists have been tracking the evolution of both since Hubble's launch.

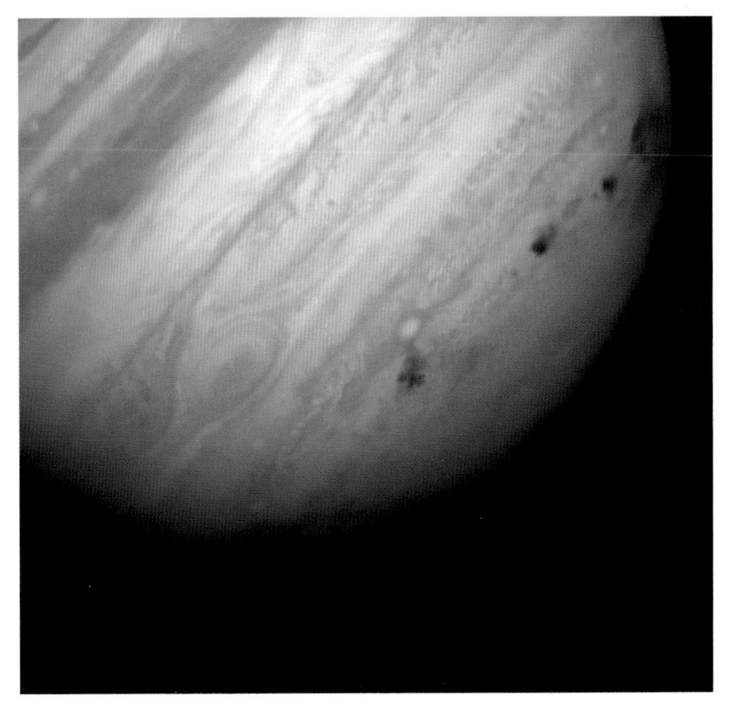

Plate 22: The impact of Comet Shoemaker-Levy into Jupiter

When the remains of Comet Shoemaker-Levy crashed into Jupiter in July 1994, Hubble took numerous images, showing a string of black impact sites that skirted the planet south of the Giant Red Spot and quickly evolved over days.

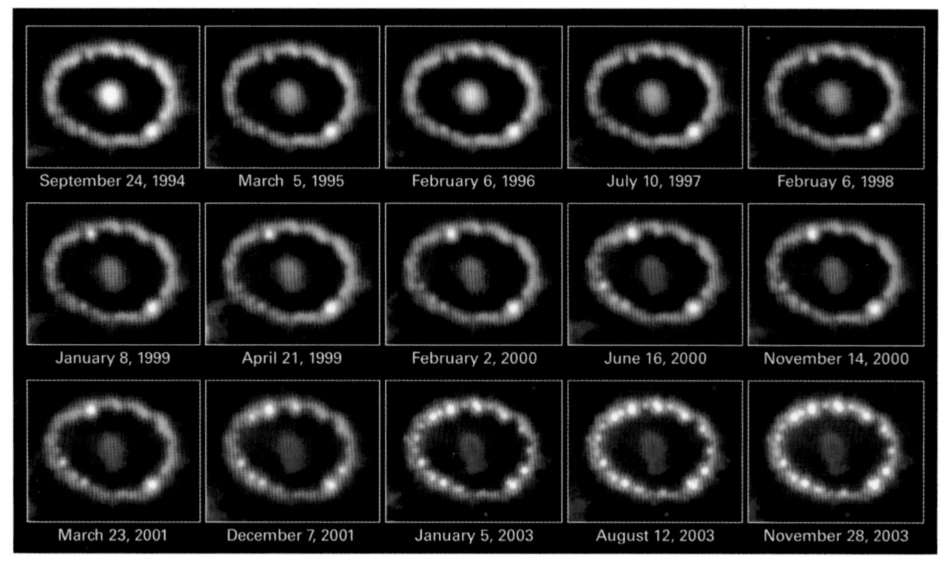

Plate 23: Changes in SN1987a over a decade

The 1987 supernova was the first visible to the naked eye since the development of the telescope. As soon as Hubble was launched in 1990 astronomers began a careful program with the space telescope to track the supernova over time. The ring, left over from a previous eruption of the star some 20 thousand years ago, steadily brightened as the outermost blast wave from the supernova began to impact it. Meanwhile, in the ring's center Hubble has been able to track the expansion of a dumbbell-shaped central structure, its two blobs expanding away from each other at about 20 million miles per hour so that twenty years after the explosion they span an area a tenth of a light-year across.

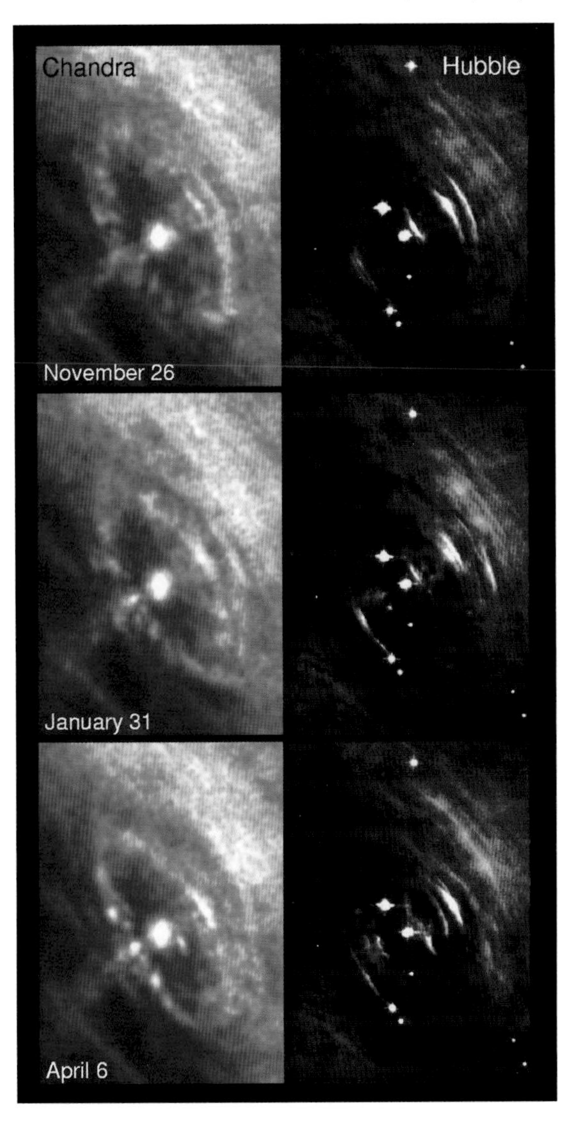

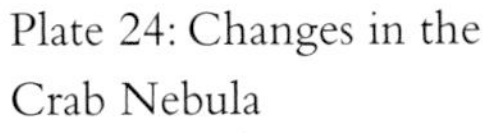
Plate 24: Changes in the Crab Nebula

In the mid-1990's astronomers used Hubble to take a series of images of the Crab Nebula over a period of months. To their astonishment they discovered that the Crab's inner nebula undergoes rapid changes. By assembling their Hubble images the scientists were even able to create a movie. This montage of the innermost section of the Crab Nebula shows a later series of observations in 2000 and 2001, using both Hubble and the Chandra X-ray Observatory. A large wisp can be seen to detach itself from the inner x-ray ring shown in all three Chandra images, then expand outward at half the speed of light into the surrounding nebula. At the same time blobs of material fly outward along a jet that points in a direction perpendicular to the rings.

Plate 25: December 1993 image of Eta Carinae

This is the December 1993 image of Eta Carinae, taken immediately after Hubble's repair. The white color of the two lobes is thought to be dusty material reflecting starlight, while the surrounding red cloud is made up of nitrogen gas. Later Hubble images have shown changes in the two white lobes as they expand outward at approximately 1.5 million miles per hour.

Recent research has indicated that Geoffrey Burbidge's 1962 speculations about Eta Carinae were correct and that the star is one of the most likely stars in the Milky Way to be the next supernova. In fact, recent data even suggest that Eta Carinae could go supernova at any time. As noted in 2007 by a scientific team led by Nathan Smith of the University of California at Berkeley, "We had better keep a watchful eye on Eta Carinae." (Smith et al. 12) with a precise number. Ground-based telescopes did not have the resolution to see Cepheids in more than a handful of galaxies, none more than one or two hundred million light-years away. Other tools to measure distance, such as the brightness of galaxies, were simply not reliable or precise enough. As a result, estimates of the age of the universe in the late 1980s and early 1990s ranged anywhere from 12 to 20 billion years. To more precisely measure the expansion rate, which astronomers dubbed the Hubble Constant in honor of Edwin Hubble, required a significant increase in resolution, which would allow the measurements of Cepheids in galaxies far more distant.

In other words, measuring the precise age of the universe required a telescope above the atmosphere. When it was launched, one of the Hubble Space Telescope's primary missions was to answer this question. In fact, it had been the main reason the astronomers had resisted shrinking the telescope's mirror below 2.4 meters. Smaller than that and the space telescope would not have been able to see Cepheids far enough away to resolve the issue.

Once the space telescope was fixed, astronomers were able to use it to identify Cepheid variables in almost three dozen different galaxies in the distant Virgo, Fornax, and Leo clusters of galaxies, and were able to narrow the range of estimates for the rate of expansion to about ten percent. From this, they were able to estimate that the Big Bang occurred approximately 13.7 billion years ago, give or take about a billion and a half years.[14]

But that was not all. Of all of the discoveries that have come with the help of the Hubble Space Telescope, the one that is probably the most significant is the one that no scientist expected: this expansion of the universe seems to be *accelerating*, on vast scales.

Before the mid-1990s, astronomers wondered whether the mass of the universe was sufficient to slow the expansion enough so that it would eventually stop growing and collapse back onto itself. If so, the universe would be considered "closed." If not, the universe was "open," and the expansion would continue forever, until the distance between our galaxy and all its neighbors was so great that we could no longer see them. In either case, astronomers assumed that the expansion was being slowed by gravity. Just as gravity will cause the debris from an explosion to slow down as it flies away, the mass of the galaxies exerts a force on the expansion, causing it to slow as well.

In the mid-1990s two different teams of astronomers, the High-z Supernova Search Team, led by Brian Schmidt of the Siding Spring Observatory in Australia, and the Supernova Cosmology Project, led by Saul Perlmutter of the Lawrence Berkeley National Laboratory in California, used a host of ground-based telescopes as well as Hubble to search out as many distant type Ia supernovae as possible. Since the intrinsic brightness of these supernovae can be determined, astronomers can estimate their distance with reasonable accuracy, depending on their apparent brightness.

To the astonishment of the scientists on both teams, at greater distances the supernovae were dimmer than expected, by 25 percent, meaning that they were farther away than expected and implying that over the last five billion years the expansion rate was actually *increasing*. It was as if the pump forcing air into the balloon began working faster as the balloon got larger.

When shortly after this discovery I asked Mario Livio, a scientist at the Space Telescope Science Institute, his thoughts on the matter, he couldn't help noting his amazement. "This phenomenon was exactly the *opposite* of what we would have expected."

Consequently, astronomers have attributed the acceleration to something they have labeled "dark energy" (in that they as yet have no idea what it is), and have estimated that it somehow acts to repulse the effect of gravity over large distances. Furthermore, scientists think this mysterious dark energy constitutes almost two-thirds of the universe.

And most amazing of all, Einstein himself had both predicted and rejected its existence. In the 1917 he had inserted into his equations something he dubbed the Cosmological Constant in order to make those equations describe a static universe. When Edwin Hubble later showed that the universe was not static but expanding, Einstein renounced this constant, eventually calling it his "biggest blunder," according to physicist George Gamow. Eight decades later, data from the Hubble Space Telescope proved that Einstein had actually been right, and that a variant of his cosmological constant was required to make his equations work.

Once again the universe has surprised us. As the mathematical biologist J.B.S. Haldane once wrote, "The universe is not only queerer than we suppose, but queerer than we *can* suppose."[15]

The Hubble Space Telescope's contribution to this particular and astonishing discovery was important, though not central. Many other telescopes were involved and were in fact more important, such as the 10-meter Keck telescopes on Mauna Kea in Hawaii and the 4-meter Cerro Tololo Inter-American Observatory in Chile. These and other ground-based telescopes were often more available or technically more appropriate for obtaining the precise data that were necessary.

For astronomers, the space telescope's most significant contribution to the field of cosmology instead came from its deep field images, beginning with the Hubble Deep Field in 1995 and followed by a second Deep Field in 1998 and the Ultra Deep Field in 2004. (See color plates 14, 15, 16, and 17.) Here, Hubble looked where no other telescope could look, and saw things that no other telescope had ever seen.

The first Hubble Deep Field, taken in December 1995, was the inspiration of Robert Williams, director of the Space Telescope Science Institute after Riccardo Giacconi left in 1992. As director Williams had a certain amount of discretionary observing time available to him. Although his own astronomical specialty involved planetary nebulae and the shells of novae and supernovae, he was intrigued by the possibilities of having Hubble look as far and as deep as it could. "We would not only be looking very far away," he explained in 1998, "but we would also be looking back very far in time."

Williams decided to devote ten days and 150 orbits to imaging what appeared to be a blank spot in the sky; he also decided to make the image immediately available to the entire astronomical community. Unlike most other observing programs, he would not hold onto his one year of proprietary rights.

What Hubble saw in this first image turned out to be mind-boggling, producing an image packed with more than 3,000 distant and strangely shaped galaxies from the universe's early history. "An awful lot of galaxies," Roger Blandford, then at Caltech, remembered thinking in surprise when he first saw it.

And like children at a candy store, the world's astronomers crowded around the picture, writing hundreds of papers about it, shedding light on such topics as the history of star formation, the early evolution of galaxies, and the development of today's vast galactic clusters. From this image as well as subsequent deep fields astronomers were able to begin outlining the primeval formation of galaxies, including what they labeled as a "baby boom" of star formation that occurred when the universe was approximately a fifth of its current age.

Before Hubble, this early universe had been inaccessible. Now the scientists could reach out and at least get a glimpse of these distant and faint objects and how their evolution over time produced the galaxies we see today in the local universe. As Richard Ellis, then of the University of Cambridge, noted, "It is hard to remember an image that has had such an impact in such a short time."[16]

■ ■ ■

For this writer, however, none of Hubble's observations have been as significant as one that the public relations people at NASA refused to include in its initial press release for the American Astronomical Society meeting on January 14, 1994. From my perspective as a historian, this image is significant not so much for its scientific uniqueness as for the impact it has had on both society and the culture of astronomy.

Just prior to the special session at that 1994 AAS conference, NASA held a press conference where the space agency unveiled for reporters WF/PC-2's spectacular images of Orion, M100, and 30 Doradus, and COSTAR's images of Supernova 1987a and Nova Cygni 1992. All of these images were remarkable and exciting. The reporters, including myself, couldn't help being as enthused and electrified as the scientists, seeing without doubt that Hubble had been repaired.

Still, these images didn't show us much we hadn't seen before. M100 still looked like a galaxy, even if the resolution was finer than ever before. The Orion Nebula still looked like a nebula, though the proplyds were cool. And though the ring surrounding the 1987 supernova was new, it pretty much resembled a host of previously photographed planetary nebulae.

Following the official press conference and the presentations by the WF/PC-2 and COSTAR teams, I wandered down to a second ballroom. Here, the astronomers had set up what they called their poster sessions, each poster illustrating a particular research topic with the astronomer in charge standing by to answer any questions about his work. Hundreds of scientists and reporters roamed down aisle after aisle of these temporary display panels, reading and chatting with other scientists about the posters tacked up on the panels. As was to be expected, the displays included many complex formulae, pretty illustrations, and the typically fuzzy images of stars, planetary nebulae, and galaxies taken by the ground-based telescopes of the time.

No one strolled past one display, however, and in fact it had drawn a rather large and fascinated crowd. The poster showed the WF/PC-2 image of Eta Carinae, and was being hosted by Jeff Hester. Unlike the viewgraph that he had used in his main presentation only a few minutes earlier, the poster was a very high resolution print of WF/PC-2's image of the star.

To me, the picture was mind-reeling. Unlike every previous photograph taken of nebulae by ground-based telescopes, this image was sharp and clear. And it was unmistakable what it revealed. (See color plate 25.)

Here was a star exploding. In reds and blues and grays, the picture showed what appeared to be two expanding bubbles of material flying outward at great speed in opposite directions from the star. It also showed several jets, and a flat disk that seemed to cut through the star's equator.

The image was so sharp that it showed many tiny and spectacular details within it. For example, when I asked Hester how large the two big expanding bubbles were, he pointed out one of the small cloudlike patches inside the nearest bubble and noted, "That's about the size of our solar system." For the next ten minutes he and another astronomer began arguing about the expanding lobes. Were they filled gas clouds expanding outward like jets from the central star? Or were they empty bubbles, shockwaves pushing out against the nebula and leaving an empty space behind?

More than any of the other pictures at the press conference, this image illustrated how necessary it had been to fix Hubble. As good as the earlier deconvoluted image of Eta Carinae had been, it hadn't been good enough. For example, the "ladder" was now gone. It had never existed, a mere artifact of the out-of-focus mirror, the software manipulations, and an overexposed star in the center of the image. Similarly, the new image unmistakably showed that the two lobes were the result of a bipolar flow pushing outward from the star in opposite directions, not an oblate shell pinched in the center. The image's clarity ended the debate over the shape of the Homunculus. It was so clear it allowed the astronomers to finally argue about real issues, even as I watched. The star had exploded, but the large bubbles from the poles as well as the equatorial disk and jets suggested it had apparently done so more than once, for different reasons.

Yet, this image of Eta Carinae had not been included in the NASA press presentation or its official press release. When the scientists and NASA officials had been putting together that press conference, one of the public affairs officials at the Space Telescope Science Institute had rejected the Eta Carinae image as too strange. As Hester remembered, "[It] did not pass the grandmother test." According to this test, "if you showed that image to your grandmother she would not immediately recognize that it was an image of an astronomical object." While the other pictures were clearly those of a galaxy or a nebula or a star, Hester said, "Eta Carinae was judged to be not sufficiently astronomical to be part of the press release."[17]

Ironically, in subsequent years it is this image of Eta Carinae that has become one of Hubble's most famous. And in this sense at least it possibly represents Hubble's most important revelation, not so much for its scientific import—which isn't insignificant—but for the cultural impact Hubble has had on the general public and astronomy. With this image, it was eminently clear that for the first time, the human race had put on a pair of good eyeglasses and could finally see the heavens *as they really are*. The image was so clear that ordinary citizens could look at it and instantly understand what they were seeing.

The consequences of this clarity have been a passionate increase in the public's interest in astronomy, drawing millions of people into the science in a manner that was unprecedented. In classrooms throughout the world are plastered Hubble pictures, giving children a detailed and exciting introduction into the world of science and what it can teach us about the universe. Nor are the kids the only ones entranced. Hubble images are eagerly viewed by millions of people almost immediately upon release.

In turn, this increased public interest in astronomy has led to a significant increase in the 1990s and 2000s in funding for the construction of bigger and better telescopes, both in space and on the ground. Besides space telescopes such as the Chandra X-ray Observatory, the Spitzer Infrared Observatory, and the James Webb Space Telescope, there has grown a whole city of bigger and better telescopes on top of Mauna Kea in Hawaii in the last fifteen years, including the twin 10-meter Keck telescopes and the 8.3-meter Subaru and 8.1-meter Gemini telescopes. Then there are the numerous new telescopes built across the world. For example, the Very Large Telescope in Chile is made up of four 8-meter telescopes arranged in an array that produces the equivalent of a 16-meter mirror; this is more than three times larger than the 200-inch Hale Telescope at Palomar Mountain and almost seven times larger than Hubble. And though it is presently the largest in operation, the Very Large Telescope is only one of many. A European group was even considering building a ground telescope, appropriately called the Overwhelmingly Large Telescope, with an aperture as big as 100 meters across. They have since renamed this telescope the Extremely Large Telescope, and are aiming for an aperture from 40 to 60 meters across.

Nor has Hubble's clarity only brought astronomers the reward of increased funding. Though the optical wavelengths form only a small part of the tapestry astronomers use to try to understand what is going on in the heavens, to all humans the optical wavelengths give us a direct window into those phenomena. As Riccardo Giacconi noted, "The images themselves seem to convey information to people through the visual means, and this information is essentially correct."[18]

Consider Eta Carinae again. Since the day John Herschel had seen Eta Carinae suddenly burst forth in 1837, astronomers had struggled to figure out the shape and structure of its Homunculus. This struggle had been characteristic of almost all astronomical research, limited as it was by the cloudy, hazy, fuzzy, and unclear images produced by ground-based telescopes combined with astronomy's restricted access to a large portion of the electromagnetic spectrum. Moreover, though astronomers had been able to construct detailed theories about what those earlier images and data were showing them, it was challenging if not impossible for them to determine which of those theories correctly interpreted what was there.

After Hubble, however, what had been considered the best images of the nebula by Gaviola in 1945 were now merely out-of-focus blurs. A "little man" Eta Car certainly wasn't. Here, at last, was a clear, unequivocal image. Here was truth, staring us all in the face. With this image astronomers no longer had to argue about *what* they were seeing. Now their job was to try and figure out the details of what was happening from clear images of the event.

Meanwhile, for the general public, this Eta Carinae image was only the beginning. In the years since, Hubble's outpouring of spectacular images has been breathtaking and continuous. And with each new image the public has been given a better view of its place in the universe.

It was as if the human race had been living in a fog, and that fog was suddenly lifted, revealing the heavens in all their glory.

■ ■ ■

By 2003, there was little indication that the dazzling stream of data from Hubble was about to end. A fifth shuttle servicing mission was scheduled to occur sometime in mid-2004, giving Hubble new gyros and batteries so that it could continue to operate for years to come. Moreover, two new instruments, the Cosmic Origins Spectrograph (COS) and Wide Field Camera 3 (WFC-3), had been built and were in storage at the Goddard Space Flight Center, awaiting launch. WFC-3, built for $83 million, was to replace the aging WF/PC-2 with two 2048 × 4096 CCDs, more than ten times larger than the original CCDs installed on Hubble's first camera. COS, meanwhile, was intended as a supplement to the spectrographic capabilities of STIS. Costing $65 million, it was ten to thirty times more sensitive than STIS.

With these improvements and repairs, astronomers expected Hubble to go on producing grand discoveries for a decade or more.

On February 1, 2003, however, the space shuttle *Columbia* was preparing to return to Earth after completing a sixteen-day science mission. The seven-person crew, including the first Israeli astronaut, had performed a wide range of scientific experiments, from measuring the changes in their own calcium levels during the flight to taking the first photographs of the electromagnetic "elves" and "sprites" produced by lightning in the upper atmosphere. They had measured dust in the air over the Mediterranean and ozone in the ozone layer. They had used an incubator to hatch a fish from its egg and a silk moth from its cocoon. They had tracked changes in their bone density, their blood pressure, their cardiovascular system. In fact, the crew had used practically every second of their entire mission working with more than seventy scientists on the ground to complete their packed research schedule.

Five minutes into the spacecraft's descent, however, controllers in Houston noticed something wrong. Within a few seconds, the readings from four temperature sensors located in the shuttle's port or left wing had been lost.

Still, everything else seemed normal. As the guidance and navigation controller told the flight director, "We have good trims. I don't see anything out of the ordinary."

Less than five minutes later, however, another ground controller noted the loss of tire pressure on both port tires, still folded inside the shuttle's left wing.

The news was quickly relayed to *Columbia.* Commander Rick Husband responded, "Roger," but his transmission was cut off mid-word.

At that moment, video footage taken on the ground suddenly showed the spacecraft engulfed in bright flashes.

The space shuttle *Columbia* had just broken up.

7

Abandonment

Thanksgiving weekend, 2003. The time had come for NASA administrator Sean O'Keefe to make a decision. As happened every year since the space agency's inception, after months of meetings and negotiations the annual budget cycle was drawing to a close. In only a few weeks he would have to finalize the 2005 NASA budget for presentation to Congress, which required him now to make some decisions on where to spend that money.

This particular budget cycle, however, had been especially difficult. Since the loss of *Columbia* ten months earlier, the space agency had been faced with an avalanche of criticism. The worst of this came when the official investigation into the accident released its findings. Given the job of finding out what happened to destroy the orbiter, the members of the Columbia Accident Investigation Board (CAIB) had decided to dig far deeper. Their conclusions, released in August 2003, not only described in detail the technical failures that caused *Columbia* to break up but also found long-term and systemic failures within NASA's entire management. There had been an inability to face squarely the engineering problems that caused both the *Columbia* and *Challenger* accidents, but also a failure to build a replacement for the shuttle as well as properly manage the entire U.S. manned-spaceflight program.

The CAIB's report was harsh and damning. In investigating the *Columbia* accident, it noted how NASA's management style was beset by arrogance and overconfidence, preventing managers from responding to safety warnings or technical problems with an open mind. As their report noted, "the NASA human spaceflight culture manifested . . . in particular a self-confidence about NASA possessing a unique knowledge about how to safely launch people into space," a self-confidence not reflected by reality. Instead, the investigation found NASA to be a place with a "cumbersome organizational structure, chronic understaffing and poor management principles." Rather than use their brains and common sense, NASA's management too often relied on complicated and unwieldy rules for determining what action to take.

Worse, though the agency's managers demanded that their engineers adhere to stifling procedures and a rigid chain of command, making design work difficult and often impractical, these same managers often ignored the rules to make their own life easier.

The best and most horrifying example of this adherence to or misuse of protocol came during *Columbia*'s flight. As the shuttle had risen through the clear Florida sky, several cameras had captured what looked like pieces of foam breaking off from its external tank and hitting the orbiter's left wing. A number of engineers, worried that *Columbia* might have been damaged, wanted to use military surveillance cameras to photograph the orbiter while it was still in space.

Not only did NASA managers deny these requests, they showed a remarkable lack of interest in finding out if there even was a problem. "Management seemed more concerned about the staff following proper channels (even while they were themselves taking informal advice) than they were about the analysis."

The board noted how similar management practices had been cited as one of the causes of the *Challenger* accident in 1986. Then, engineers had pleaded with managers to delay the launch because they were worried that the O-rings used to seal the joints between the shuttle's solid rocket booster segments would harden in the cold weather and therefore fail to prevent the leak of hot gases. Then, the managers had dismissed these worries, and proceeded to launch with disastrous results. To the investigation board, the two incidents demonstrated how little the space agency had apparently changed in the intervening fifteen years.

Nor did the investigation board limit its harsh criticisms to NASA's management. The CAIB threw a wide net, chiding past presidents and Congress for many if not all of the management problems at NASA. For every specific issue of difficulty that NASA had faced since the 1970s—the original shuttle design, the agency's yearly budget, the structure of its shuttle program, its schedule and recent efforts to build the International Space Station and replace the shuttle—the board bluntly condemned the contribution of Congress and past presidents as "a failure of national leadership."

Above all, the board's report admonished elected officials as well as the American public for neglecting to give NASA a clear overall vision, a "lack, over the past three decades, of any national mandate providing NASA a compelling mission [for] requiring human presence in space." As a result, the board noted, "NASA remained a politicized and vulnerable agency, dependent on key political players who accepted NASA's ambitious proposals and then imposed strict budget limits."[1]

At the same time the investigation was coming to these brutal conclusions, a movement had been gathering inside the administration of President George W. Bush to address this last charge. Throughout the spring and fall of 2003 a number of Washington bureaucrats from a wide collection of cabinet agencies began having regular meetings in the executive office of the President in order to hash out a better vision for America's space program. The meetings had started with a group of lower level White House staffers and had eventually included such high-powered officials as Vice President Dick Cheney, Deputy Secretary of State Richard Armitage, and Deputy National Security Advisor Stephen Hadley. The goal was to come up with a more ambitious and coherent space program. By Thanksgiving, 2003, the work of this White House group had culminated in a complete refocusing of all of NASA's efforts toward the human exploration of the solar system, starting with a return to the Moon sometime in the next decade.[2]

When first told of these plans, O'Keefe immediately advocated a big budget increase to pay for them. To his dismay, these same White House officials refused his request, noting that under President Bush's orders domestic spending was supposed to be kept flat in 2005. After much lobbying with the White House and the Office of Management and Budget (where O'Keefe had previously been deputy director), he managed to negotiate a small increase, one billion dollars spread over the next five years.

It wasn't much, considering the ambitious nature of the emerging White House space plans. Thus, as O'Keefe was trying to finalize his budget that 2003 Thanksgiving weekend, he struggled to figure out how he was going to get it all done while simultaneously meeting all of the CAIB's strict safety recommendations.

For O'Keefe, the *Columbia* accident was a very personal tragedy. When *Columbia* broke up he had been waiting at the Kennedy Space Center's landing strip with the astronauts' families, ready to celebrate the flight's successful end. Instead, it became his job to tell everyone that their loved ones had died. Then, as the investigation moved forward over the next several months, he had presided at the memorial services for each astronaut. Again and again he had to confront the faces of spouses and children and try to explain to them why it had happened and why the loss had been worth it. As he said in testimony before the National Academy of Sciences a year later, the experience "was emblazoned in my memory like it was yesterday. There's not a day that goes by I don't think of that."[3]

This experience was not something O'Keefe was prepared for. A self-admitted Washington, DC "bean counter," he had never counted on dealing with death in his job at NASA. Trained as a manager and public administrator who had worked his way up the bureaucratic chain of command, first as a staff member of the Senate Appropriations Committee and later as comptroller and chief financial officer of the Department of Defense, O'Keefe was not the kind of person one would expect to lead an organization that was going to send astronauts to the Moon. Instead, he preferred to focus on managing organizational charts and figuring out how to save the taxpayer money.

O'Keefe also liked to use arcane bureaucratic phrases like "owning the decision," "spiral development," and "stovepiping." In his press appearances he tended to speak in a rambling manner, drifting from half thought to half thought in a kind of stream of consciousness. One journalist once described to me how this speaking style made O'Keefe "the man who could not be quoted." Not only did he not give reporters good soundbites, his manner of speech often made it impossible to pin him down on any issue.

O'Keefe's nebulous and rambling approach to public speaking was clearly reflected in his management of NASA. During his term the agency often took a very long time to come to any clear and precise decision about anything it needed to do. For example, after Bush's announcement of the new space vision O'Keefe had the agency create an

Fig. 7.1. Sean O'Keefe during a press briefing at the Kennedy Space Center immediately after the loss of the space shuttle *Columbia*, February 1, 2003. (Photo courtesy of NASA)

Advanced Planning and Integration Office. This office in turn spent most of 2004 creating thirteen Strategic Roadmap committees, each with anywhere from nineteen to thirty-four members, as well as an additional fifteen Capability Roadmap committees, each charged with outlining some of the specific technical requirements of Bush's vision.

Throughout the spring of 2005 each of these committees held meetings, talking interminably about NASA's future plans. At one point during the second meeting of the Robotic and Human Exploration Lunar Exploration Roadmap Committee I watched in disgust as more than thirty people from NASA, the aerospace industry, and the scientific community spent almost an hour discussing some minor editing changes to the *language* of their strategic roadmap plan. That nothing was actually being designed or built seemed to bother no one.[4]

O'Keefe's slow and bureaucratic management style could be seen in myriad other ways. For example, compare the actions of the U.S. and Russian space programs during 2004. Bush announced his new vision for NASA in January 2004, calling for a new manned spacecraft to replace the shuttle. Fourteen months later, all NASA had been able to accomplish was to release a vague RFP that included very few concrete design specifications but required any winning contractor to fill out an almost endless amount of paperwork on what they were doing.

During that same fourteen months the Russians not only announced their plans to build a replacement for their manned Soyuz capsule, they actually built a full-scale prototype which they began showing off at various air shows.

Another example: The space vision put together by Bush's administration required that NASA use the shuttle to complete the International Space Station, then retire the shuttle by the end of 2010. Yet, more than a year after Bush made this plan public the agency under O'Keefe had still failed to outline exactly how many shuttle missions it would take to complete the station. At any time the announced number of necessary flights mentioned by O'Keefe and other NASA officials could fluctuate anywhere from eighteen to twenty-eight, with little clarity about what this schedule would accomplish when.

It is therefore not surprising that a bureaucrat like O'Keefe was ill-prepared on February 1, 2003, to face the loved ones of seven dead astronauts, killed just minutes before they were to return home.

As O'Keefe pondered his budget that 2003 Thanksgiving weekend, one budget item stood out that might give him some breathing room if he eliminated it. In addition to the two dozen or so remaining shuttle flights to ISS, the flight manifest included the fifth and last servicing mission to the Hubble Space Telescope.

By 2004 the telescope had been in orbit for nearly fifteen years. Its batteries, what one engineer once told me "were the world's best rechargeable batteries ever built," had been manufactured in the early 1980s and were now almost a quarter-century old. They had been charged and recharged more than 75,000 times since launch, and were showing a slow decline in their recharge capacity. Based on this rate of decline, sometime in 2009 they were expected to no longer be able to power the telescope.

Then there were Hubble's gyros. Three had been replaced in 1993, followed by two more in 1997. Then four had failed in 1999, forcing the separate repair mission in that year to replace all six. Even so, two more had to be replaced during the 2002 servicing mission.

Despite upgrades to later gyros, engineers had not yet figured out how to make Hubble's gyros last more than a few years. By 2004 two of these replacement gyros had failed, with a third showing signs of wear. Since Hubble needed at least two gyros to be able to do any science at all, and three to take its sharpest pictures, the telescope no longer had much margin should another gyro fail. In order to keep the telescope operating past 2008 it was imperative that it get a new set of gyros.[5]

The telescope had other maintenance concerns. Some of its thermal insulation needed replacing. One of its fine guidance sensors needed to be replaced. A new cooling system had been built that could be attached to the telescope's aft shroud and help radiate excess heat outward. A backup data handling unit was ready to provide the telescope some more computer redundancy.

In addition to these routine maintenance tasks, NASA had also paid for and built the two new instruments, the Wide Field Camera 3 and the Cosmic Origins Spectrograph (COS), now sitting at Goddard and awaiting launch.

To O'Keefe, however, improving and maintaining Hubble was his least worry. In responding to the CAIB's conclusions, O'Keefe had decided to do more. As he stated in testimony to Congress only days after the release of the CAIB report, "I can assure you that we will not only implement the CAIB's recommendations to the best of our ability, but we are also seeking ways to go beyond their recommendations."[6]

In the months following *Columbia*'s destruction, however, he had found this commitment increasingly difficult to meet. Day by day there were new unexpected engineering problems. For example, the investigation had recommended that before another shuttle flew NASA should develop some method for repairing the shuttle's thermal protection system should it be damaged again during launch. Accomplishing this recommendation was not simple, however. Most known adhesives did not work very well in the vacuum of space. On top of this, no matter what

Fig. 7.2. Sean O'Keefe inspects some of the recovered wreckage from *Columbia*, April 28, 2003. (Photo courtesy of NASA)

repair system the engineers proposed, there was no way to find out if it worked without first testing it in orbit. Such in-orbit testing, however, required at least one or two shuttle flights to fly without a working repair system, a circumstance that broke O'Keefe's promise to meet all of the board's recommendations before he put another shuttle into orbit.

Then there was the problem of inspecting a shuttle for damage while it was still in orbit. It was a relatively straightforward procedure to give the shuttle a thorough inspection when it went to the International Space Station. Before docking the shuttle could fly in formation with the station so that the station's astronauts could photograph and inspect its underside. Then, once docked, closer inspection by spacewalk or with the station's robot arms was also relatively manageable. Should these inspections discover any serious damage, the station also provided time to mount a shuttle rescue mission. Station supplies could sustain a shuttle crew for anywhere from one to three months while a second shuttle was prepared for launch.

A mission to Hubble had none of these luxuries. Inspecting the bottom of a shuttle sent to Hubble would be difficult and complicated. NASA's engineers had been working hard to develop an extension boom for the shuttle's robot arm that could reach below the shuttle and position a camera there to photograph its underbelly. However, that work was going slowly. Engineers didn't know if they would be able to provide the camera enough light. They didn't know if they could get sufficient resolution to see very small holes or cracks. And they weren't sure if a Hubble shuttle mission would have enough time to do a thorough inspection.

Even if these obstacles were overcome and the astronauts were able to properly inspect the shuttle, a rescue mission would still be required if they found any damage. A shuttle's onboard supplies, however, were no match for those at ISS. The most a shuttle's life support systems could be stretched was about a month, barely enough time to get a second shuttle prepped and launched.

In reviewing these facts, O'Keefe found himself inclined to delete the Hubble mission from his 2005 budget. If he left it out he would save money and also eliminate what he considered a very risky mission from his schedule.[7]

If O'Keefe had had any doubts about this decision, he found these doubts eased by the very recommendations that astronomers had been making to NASA over the last decade. In fact, a superficial look at these recommendations gave the unfortunate impression that the very people who had been using Hubble throughout the 1990s to revolutionize astronomy had apparently little enthusiasm for keeping the telescope operating past its nominal fifteen-year mission.

In the 1980s, before Hubble was launched, the assumption had been that when the space telescope's mission ended in the mid-2000s it would be replaced with a much larger optical telescope. For example, in 1985 Pierre Bely at the Space Telescope Science Institute proposed a Super Space Telescope with an increase in resolution five to ten times that of Hubble. As he noted, "free of atmospheric turbulence, a [super space telescope] would be king in the high resolution game."[8]

Bely's letter proposal led to a joint workshop by the institute and NASA and held at the institute in September 1989, only seven months before launch. Dubbed "The Next Generation Space Telescope," the meeting's sole purpose was to begin the process of replacing Hubble. The scientists at the conference proposed a whole range of ideas, almost all of which centered on building a better and more advanced version of Hubble.

For example, Aden and Marjorie Meinel were back, this time with a proposal for an ambitious lunar telescope, built in conjunction with a lunar outpost and made up of an array of smaller telescopes that would work together to produce a single image. Also there were some of the same engineers from the Marshall Space Flight Center who had built Hubble, giving presentations outlining a range of designs with mirrors from 8 to 1000 meters in diameter. Other proposals included a space telescope whose 20-meter mirror would deploy in orbit, a mirror made up of segments that would be assembled in orbit, and an orbiting array of telescopes that would work together.

All these designs assumed the construction of a telescope operating in optical, ultraviolet, and infrared wavelengths. As the workshop's conclusions pointed out, a telescope with such capabilities would be "vital for the study of the most fundamental questions of astrophysics."[9]

By the mid-1990s, however, the astronomical community was less inclined to build another optical telescope, for many justifiable scientific, technical, and economic reasons. Instead, they recommended that Hubble be replaced with a space-based telescope optimized for infrared observations, with no optical capabilities at all.[10]

This change in thinking took place with the writing of the 1996 Dressler report (officially called "HST and Beyond"). After the 1993 repair of Hubble, AURA—the university consortium that ran the Space Telescope Science Institute—put together a committee to look at what the next generation space telescope should do. Chaired by Alan Dressler of the Observatories of the Carnegie Institution in Pasadena, California, and including many of the most prominent astronomers in the world, this committee began its deliberations focusing on "space-based optical/UV/near-IR astronomy for the period 2005–2020."[11]

By the time it published its report in 1996, however, the Dressler committee had deleted the optical and ultraviolet wavelengths from its next generation space telescope recommendations and instead proposed a Hubble replacement devoted solely to infrared wavelengths. In polling many astronomers, the committee learned that a large number believed that a coming new generation of ground-based giant telescopes was going to make a new space-based optical telescope unnecessary. New computer and laser technologies, called adaptive optics, were expected to be able to cancel out the unsteadiness of the Earth's atmosphere. "By 2005," wrote committee member Peter Stockman of the institute, "ground-based telescopes may have surpassed the resolution of [Hubble] and a space 4 meter [telescope] at visible wavelengths."

The committee also recognized that the cost of developing a larger optical/ultraviolet space telescope was daunting. The short wavelengths of optical and ultraviolet light require very precise optics. Though the technology for polishing and assembling gigantic mirrors far larger than Hubble's was becoming practical, getting such a large mirror into orbit was expensive and difficult. Moreover, developing the capability of pointing such a large telescope while in orbit was neither simple nor easy. "It would have cost an enormous amount of money," Weiler explained. "The mirror tolerances drive the costs way up." Moreover, at the time the Dressler committee was beginning its work, Hubble's spherical aberration problems had only recently been fixed. It seemed too soon politically to argue for an even more expensive and technically more difficult optical telescope.

Instead, NASA officials such as then administrator Dan Goldin and Ed Weiler, who was an active member on the Dressler committee, encouraged the astronomers to come up with something that could be built for a cost similar to Hubble's but that would have the ability to do something really different and unique. "[Weiler] said that the best way for us to sell a new mission was to have an objective," explained Stockman. "You must have a key objective."

At the same time, the deep space cosmologists (many of whom were also members of the committee) were lobbying for a high-resolution infrared space telescope to probe the deepest regions of space. The scientific possibilities in the infrared seemed significant. Such a telescope would allow astronomers to look back farther than any telescope had ever seen, giving them their first glimpse of the first stars, the first black holes, the first galaxies, the first of everything. "We thought the fascinating science would be in the high redshift universe," remembered Stockman. To do such Big Bang research at the highest redshifts, however, was impossible for optical telescopes, either on the ground or in space.

At high redshifts optical light would have been shifted into the infrared, where an optical telescope simply couldn't see it.[12]

In addition, the technical problems of building a four-meter infrared space telescope were far less challenging, and therefore less costly, than that of an optical/ultraviolet telescope. A 4-meter infrared telescope would also be five times larger than the largest infrared telescope then being planned (what has since become the Spitzer Infrared Observatory). This increase in capability dwarfed the less significant improvement that would have been achieved in the optical/ultraviolet by going from Hubble's 2.4 meters to 4 meters. This was also one of the reasons that, though there were those in the astronomical community who were advocating the construction of a new and larger optical/ultraviolet space telescope to replace Hubble, their arguments were not convincing to the majority of astronomers. To most astronomers the research payoffs in the optical/ultraviolet just did not seem as interesting or as revolutionary.

Thus, though the Dressler committee recommended extending Hubble's life for five years in order to make up for the years lost due to spherical aberration as well as to keep it operating until a next-generation infrared telescope could be launched, it also recommended "a much more economical style of operation beyond 2005 . . . with no servicing or instrument replacements after 2005, a final boost into a higher, long-lived orbit, and a possibly reduced instrument complement and limited modes of operation." Hubble's replacement in turn would operate in the near- to mid-infrared and not in the optical wavelengths. "The Committee recommends a large infrared-optimized observatory as the highest priority post–HST facility–class mission for NASA."

As Steve Beckwith, director of the Space Telescope Science Institute after Robert Williams, explained to me in 2004, the astronomical community had made a calculated strategic decision to get the infrared telescope built first, then deal with getting a replacement for Hubble. "You have to prioritize. . . . HST was up there, it was going to be supported for at least another decade. . . . What [the astronomers] wanted to do was to push [the next-generation infrared space telescope, making it] the key number one priority."[13]

Instigated by the very organization that operated Hubble, the Dressler report had great influence, and by 1998 the die was cast. In that year the European Space Agency held a three-day conference in Liège, Belgium, on the kind of telescope to build after Hubble. In contrast to the goal of the conference in 1989, the plan was now to make the next space telescope an infrared instrument, optimized to study deep space cosmology.[14]

This was followed in 2001 by the "Astronomy and Astrophysics in the New Millennium" report, written by some of the most important astronomers in the world. Similar in nature to the Greenstein decadal report in 1972 and the Bahcall decadal report in 1991, this new decadal survey was especially blunt about what astronomers wanted to do in space after Hubble. "The committee has not recommended any new moderate or major missions for space-based [ultraviolet] or optical astronomy for this decade." Instead, this decadal survey once again recommended the completion of one final servicing mission to Hubble, followed by "reduced operating costs." Meanwhile, the Next Generation Space Telescope, which we now call the James Webb Space Telescope (JWST), would devote itself to infrared wavelengths, while large ground-based telescopes using adaptive optics would handle optical wavelengths. Ultraviolet research, which according to the survey "is impossible to observe . . . from the ground," was provided no replacement after Hubble. The survey even conceded that once Hubble was de-orbited such research would be impossible, "for this decade at least."[15]

To an outsider such as Sean O'Keefe, unaware of the detailed scientific considerations behind these decisions, this willingness of astronomers to abandon space-based optical and ultraviolet astronomy was deadly. Even as Hubble was allowing astronomers to roll out revelation after amazing revelation, on the surface it seemed as if in the future they were quite willing to live without an optical telescope.

In fact, it wasn't until *after* the *Columbia* shuttle accident that a committee of scientists, headed by Bahcall, finally declared that they wanted to keep Hubble operating past 2010. Released in August 2003—the same month the CAIB report was released—not only did this Bahcall committee report state that the scheduled servicing mission to Hubble was necessary, it called for NASA to consider flying an *additional* servicing mission, in order to keep Hubble operating into the late 2010s. "The committee prefers a shuttle mission for this purpose because the same mission will also improve the performance of HST before it is brought down to Earth."[16]

Unfortunately, this last Bahcall committee report was too little too late.

There is an old saying: Be careful what you wish for, you might get it. Regardless of their sincere and thoughtful effort to maximize astronomical research across as many wavelengths as possible for the least cost, by not pushing for an optical/ultraviolet space telescope to replace Hubble in 2010, and by accepting no servicing and limited operations for Hubble after 2005, astronomers had left themselves vulnerable to having Hubble's last servicing mission canceled. Consider how it must have looked to Sean O'Keefe that 2003 Thanksgiving: if astronomers didn't think it was necessary to service Hubble after 2005, why should it matter if servicing ended now? And why should astronomers care anyway, since they didn't seem interested in building an optical telescope to replace Hubble?

As result, the astronomical community, the people for whom the loss of Hubble mattered the most, gave Sean O'Keefe the very ammunition he needed to end Hubble sooner than necessary.

Nonetheless, for O'Keefe the most important considerations remained the issues of risk and safety. A Hubble mission lacked the "safe haven" provided by ISS. A Hubble mission required sophisticated and as yet unproven inspection techniques that were easy to do at ISS. A Hubble mission required new and untested tile repair technology that at ISS could be bypassed because there was time to launch a rescue mission. From O'Keefe's perspective, these factors made keeping the Hubble servicing mission on the schedule irresponsible. Moreover, by canceling the mission he would simplify the budget process. Though the cost of doing a shuttle mission to Hubble was not a major issue, getting rid of it made it easier for this self-admitted "bean counter" to deal with his other financial challenges.[17]

Just as in 1976 when Noel Hinners had decided to delete any funding for a Hubble new start, O'Keefe intended to keep his decision to delete the servicing mission secret until late January when the entire federal budget proposal was announced by the White House.

And just as in 1976, the news leaked early. By early January, after weeks of sending out trial balloons to the press about the president's new space vision, the White House was finally ready to unveil it. On Jan. 14, 2004, President Bush arrived at NASA headquarters for a gala press conference and—with more than a dozen active and retired astronauts in the audience—called for the United States and NASA "to extend a human presence across our solar system. . . . [to] build new ships to carry man forward into the universe, to gain a new foothold on the Moon, and to prepare for new journeys to worlds beyond our own."[18] The vision set new and finite goals for the remaining space shuttle fleet. Once the shuttle fleet returned to flight, it would complete construction of the International Space Station and then be retired in 2010. To replace the shuttle, NASA would develop a new crew exploration vehicle, what is now called Orion, with unmanned test flights before 2010 and manned missions in 2014. Orion would then be used to establish a base on the Moon—as early as 2015 and no later than 2020—followed by later manned missions to Mars and beyond.

Whether one agreed with it or not, this long-term blueprint was undeniably the clearest and most focused vision for NASA since President John F. Kennedy's Moon landing commitment in 1961.

In her front page article the next day for the *Washington Post*, Kathy Sawyer added one extra detail about the plan that hadn't been mentioned by Bush or anyone else: "There will be no further servicing missions of the Hubble Space Telescope." Though none of NASA's press releases had mentioned this fact, she had picked it up from her sources in Congress, who had in turn heard about it during a budget briefing for the House Science Committee.[19]

Caught off guard by this leak and the resulting storm of emails and phone calls from scientists and members of Congress wanting to know if it was true, O'Keefe had to move fast. He hurriedly arranged a meeting for Friday, January 16, at the Goddard Space Flight Center north of Washington, DC, to officially announce the bad news to those who worked on Hubble. Before a crowd of more than a hundred people from the institute, Goddard, NASA headquarters, and many Hubble contractors, O'Keefe spoke and answered questions for almost ninety minutes.

O'Keefe was clear. The decision to cancel the last shuttle mission to Hubble had been his alone, and he wanted to tell everyone involved in person, face-to-face. In his typically rambling style he explained that the decision was not due to any single factor, but the weight of many issues, including safety and budget. He also explained that since the shuttle servicing mission would only extend Hubble's life for a short period, it was clearly expendable. Moreover, since astronomers had access to many other ground-based and space-based telescopes as well as new adaptive optics techniques, it didn't seem reasonable to him to waste NASA's slim resources trying to plan this risky mission when the agency had so many other demands on its plate.

As institute director Steve Beckwith noted in an email to his staff that day, "The mood in the room was decidedly somber."[20]

The public response in turn was, to put it mildly, loud. Very soon, NASA was getting four hundred emails a day in protest, and over the next few weeks editorials and op-eds in dozens of newspapers across the United States came out against the decision. Almost instantly several grassroots organizations started letter-writing campaigns and petitions protesting the decision. On one webpage almost 40,000 signatures were gathered in less than two months. As the *Denver Post* noted in an editorial on March 22, 2004, "The public and Congress are letting NASA know they understand the real value of the Hubble Space Telescope—even if NASA doesn't."[21]

Meanwhile, an underground but aggressive campaign to save Hubble was gearing up at the Space Telescope Science Institute, under the leadership of Beckwith. Like O'Dell in the 1970s, Beckwith was not legally permitted to lobby Congress or the public. Like O'Dell, however, he was also perfectly positioned to manipulate the situation to Hubble's advantage. Throughout the spring and summer of 2004 he very carefully arranged to schedule Hubble press releases to maximize their value. "I wanted a cadence of stuff coming out about Hubble about once a week," Beckwith explained. "We needed to make sure that everybody understood how vital Hubble was."

In one case in particular he pushed every button he could think of to influence events. In late 2003 he had used his discretionary time as director to do what Robert Williams had done in 1995, taking Hubble's longest single exposure, what Beckwith dubbed the Ultra Deep Field. For 400 orbits and more than a million seconds spread over four months ending on January 16, 2003—coincidently the very day that O'Keefe had come to Goddard to announce the cancellation of the last servicing mission—Hubble had trained its eye on one particular and barren spot in the sky in the constellation Fornax.

Fig. 7.3. Steven Beckwith. The lapel pin he is wearing is the Hubble Space Telescope. He wore this pin every day, from the day Sean O'Keefe announced the cancellation of the shuttle servicing mission until the day Michael Griffin announced that mission's reinstatement. (Photo courtesy of Steven Beckwith)

Originally Beckwith had planned to release this spectacular image in early February. When data processing delayed its preparation until late February, he decided to delay it further in order to give the spotlight to two other discoveries. The first was announced on February 20 by astronomer Adam Riess. Riess had used Hubble to help confirm the recent discovery that the universe's expansion was accelerating rather than decelerating as expected. The second, on March 4, was a spectacular series of images of the light echo from the 2002 nova explosion of the supergiant star V838 Monocerotis. The wave of light from that explosion had progressively illuminated the shells of dust and gas that surrounded the star, all captured by Hubble's cameras and put together to form a "movie" of that illumination.

Then, on March 9, 2004, only two days before Sean O'Keefe was scheduled to testify about his budget before the Senate Appropriations Committee, Beckwith organized with great hoopla a press conference at the Space Telescope Science Institute to release his Ultra Deep Field image, getting Senator Barbara Mikulski, a passionate "Hubble-hugger," a strong opponent of O'Keefe's decision, *and* a member of that Senate Appropriations Committee, to join him in removing the curtain to unveil a picture of more than 10,000 galaxies from the very beginnings of time, squeezed into an area less than a tenth the width of the Moon.[22]

Not surprisingly, Mikulski took full advantage of the spotlight to make her thoughts known on Hubble's future. "I will continue to stand up for Hubble," she declared to loud applause. Two days later she reiterated her position to Sean O'Keefe in a fiery Senate budget hearing. At that hearing she called for an independent assessment of O'Keefe's decision, demanding that he request the formation of a National Academies panel. She also told him that he was to "take no action to stop, suspend, or terminate any contracts or employment in connection with the final servicing mission until this study is completed."[23]

Faced with this firestorm, O'Keefe could do nothing but agree. Immediately after that March 11, 2004, hearing he asked the National Academies of Sciences to put together a panel to assess Hubble's future and review his decision.

The Academies panel, given the unwieldy moniker "The Committee on the Assessment of Options for Extending the Life of the Hubble Space Telescope," included many important figures from Hubble's past. Sandy Faber was on it. So was Riccardo Giacconi. Representing the astronauts were Greg Harbaugh and Charles Bolden, both of whom had flown missions to fix Hubble. The panel also included Joe Rothenberg, who had been head of Goddard during the 1990–93 repair effort.

For Faber, the decision to serve on this panel had been easy. As soon as she heard about it she had called to volunteer. She had read the CAIB report and, with her many years of work with Hubble, felt she was particularly qualified to participate. Moreover, she had strong feelings about O'Keefe's priorities. "I was absolutely incensed that we would send what was being labeled as a 'risky shuttle' to fix a worthless space station and take the risk 25 times over when we wouldn't even send it once to save [Hubble]."[24]

This was a group of experts who had shown in the past that they were not afraid to take gambles, to attempt the difficult.

At the start, however, the panel members were hardly in agreement about what to do. The astronomers worried about the cost of fixing Hubble. Would it eat up all of NASA's astronomy research budget? If so, they would rather pass on a Hubble repair mission. The engineers and astronauts on the panel wondered about the best way to maintain Hubble, and even if it was worth it.

Complicating these deliberations was an effort coming out of Goddard to service Hubble using robots. Almost as soon as O'Keefe announced his decision to cancel the manned mission to Hubble, the same NASA engineering teams that had repaired Hubble in 1993, led by Frank Cepollina at Goddard, began pushing the idea of using robots instead. "This is about taking back our destiny," Cepollina later explained to the Hubble panel. "We need an alternative way to service Hubble."

Since Hubble's launch, Cepollina had been in charge of running every Hubble servicing mission. "He is one of the most innovative brilliant guys I've ever worked with," said Jim Crocker. "He doesn't nearly get the credit he deserves for keeping Hubble alive. If we didn't have Frank Cepollina we wouldn't have ever done COSTAR, quite frankly. His leadership is why all these missions have been successful."[25]

A Goddard engineer and manager since 1963, "Cepi" (as he is known by everyone in the aerospace industry) had always been interested in the idea of manned or robot servicing in space. In the 1970s he had helped design the satellite standardization program that NASA had put into place when the shuttle program began. Under this program, all NASA satellites were built in a modular, standardized manner, allowing shuttle astronauts to visit and repair them easily. The program's successes had included a number of satellites, among them Solar Max and the Long Duration Exposure Facility (LDEF). The in-space repair of Solar Max in 1984 was particularly spectacular, requiring two spacewalks and the first use of the shuttle's robot arm to grab an orbiting satellite. In later years Cepollina had initiated design studies on the use of robots to maintain science spacecraft placed at distant orbital positions, such as the Sun-Earth L1 and L2 points more than a million miles from Earth. At these locations it would be impossible to send human missions for many, many years.

When O'Keefe came to Goddard to announce the cancellation of the shuttle servicing mission, Cepollina immediately spoke up, suggesting to O'Keefe at that meeting that there might be other options to a manned servicing mission. "O'Keefe didn't say yes, but he also said, 'Go do it, look into it.'" Cepollina said.

As a first step, Cepollina put out a call to the aerospace industry on February 20, 2004, asking for any information that anyone could offer him on putting together an unmanned servicing mission to Hubble. "We got a lot of wild ass ideas," he remembered, including using balloons and astronauts. To his glee, he also received scads of data on a plethora of already-built and available robot arms. One arm was being developed only a few miles from Washington, DC, at the University of Maryland at College Park. Another was being tested at the Johnson Space Flight Center in Houston. The best and most useful by far, however, was being built by the Canadian Space Agency, the people who built the robot arms for all the space shuttles as well as the International Space Station. Already ready to fly, this two-armed robot, called Dextre, was space-hardened, and was intended for use on the ISS. On March 12 Cepollina presented this ammunition to O'Keefe at Goddard. O'Keefe, desperate to find some way to placate both the public and the politicians who financed NASA, leaped at the idea, and immediately approved a feasibility study.

To see if Dextre could do the job, Cepollina's team put the robot through some simulations, using a variety of identical Hubble tools and parts intended for the astronaut mission. In one test, they had several astronauts in Houston control the robot at Goddard. After these successes, it appeared the technology was sufficiently developed and reliable enough for a Hubble docking mission by late 2007.[26]

During this same time period, even as the National Academies Hubble panel was forming, the first public hint that NASA was considering a robot mission reached the press. By late May, the robot concept had become common knowledge, aided by an aggressive public relations campaign from both Goddard and NASA headquarters to boost the idea of using robots instead of humans to service Hubble.

Finally, on June 1, the agency made it official, announcing in a press release that NASA was pursuing "the feasibility of a robotic servicing mission to the Hubble Space Telescope." Noting that a robot mission was required anyway so that the space telescope could be de-orbited in a controlled manner, NASA was now ready to consider other tasks, including "installing new batteries, gyros, and possibly scientific instruments."[27] This release was as carefully timed as Beckwith's Ultra Deep Field press conference, coming out only one day before the National Academies panel was slated to begin its first set of hearings in Washington, DC. At those same hearings Cepollina along with a number of other

NASA engineers put forth the agency's argument about why a shuttle mission to Hubble made no sense, while a robot mission was doable.

Cepollina's presentation was fascinating and compelling. His presentation showed how the unmanned module would approach Hubble and use a robot arm to grapple and dock with it. Then the more complex two-armed Dextre unit would plug in new batteries and install both Wide Field Camera 3 and COS along with new gyros. Finally, the robot half of the module would undock, leaving behind de-orbit engines that would be used to bring Hubble back to Earth, burning up in a controlled descent over the Pacific Ocean.

These June 2 and 3 presentations were followed three weeks later with more hearings, this time with Sean O'Keefe himself testifying. Once again he tended to ramble. Once again, he expressed a gut-feeling aversion for any manned mission to Hubble. Once again, he strongly advocated the robot option. At the same time, he expressed a willingness to entertain an open mind, at least publicly. When one panel member asked him what he most wanted from the panel, he asked for information. "I sure would like to hear about anything that you might come across that we are simply not aware of." Yet, he also indicated he wasn't interested in any conclusions that might contradict his own position. He did not want "judgment calls, [only] facts. We're looking for the facts."[28]

As is usual for big government agency heads when they appear voluntarily at some other venue, O'Keefe arrived, immediately gave his testimony, and then left. Had he been able to listen to that day's other witnesses, however, he would have instantly gotten some of that information he wanted.

For example, Stephen Beckwith had preceded O'Keefe with a concise description of Hubble's value to scientific research, noting how it had produced scientific results at a rate as much as ten times more than any other telescope or space satellite ever launched. He also reviewed all the major fields of astronomy, from extrasolar planets to the Big Bang, showing how Hubble was the only telescope able to provide the resolution necessary for much of this research.

Also testifying that day was Claire Max, deputy director of the Center for Adaptive Optics at the University of California at Santa Cruz, who described how ground-based adaptive optics could not replace Hubble's optical capabilities. She explained that for the largest telescopes that astronomers required, adaptive optics only worked in the infrared. At optical wavelengths the cost and complexity of building the necessary computers and extremely precise deformable mirrors required for adaptive optics was simply impractical. In addition, the technology only worked if a target had a star nearby for calibration. With this limitation, adaptive optics could study no more than 30 percent of the sky. Should a supernova occur in the remaining 70 percent, ground-based telescopes would not be able to take sharp optical images of the event. Similarly, all of Hubble's deep fields would have been impossible, since those locations were specifically chosen because there was nothing there.

And even if a target star was available, adaptive optics were limited to a very small field of view, mere arc-seconds, making it ineffective at either observing larger objects or doing survey work.

Other witnesses that day reported on the difficulties of remote robot operations, noting how the time delay between the ground and orbit can make such operations very risky. One of the panel members, astronaut Charles Bolden, noted that even in real-time operations on the shuttle, some astronauts had trouble using the robot arm successfully. "Not everyone can grapple," he explained. "We have had some astronauts who have been fired because they could not grapple." The witnesses that day also noted that the use of a robot arm controlled from the ground to grapple Hubble was untested, and couldn't be tested in advance of the robot repair mission. There simply wasn't enough time. Panel member and astronaut Greg Harbaugh confirmed this, adding that the last fifteen feet of any rendezvous and docking was the most difficult. Without practice and testing it was questionable a robot mission could succeed.

The rest of the summer of 2004 was not an easy time for Sean O'Keefe and NASA. The stream of criticism about his Hubble decision refused to cease, and new problems kept popping up in the effort to get the shuttle back into space. A year after the Columbia accident report's publication only five of its fifteen recommendations had been met. Then, hurricanes Charley, Frances, and Jeanne pounded Florida, damaging several NASA buildings, including the huge Vehicle Assembly Building where the shuttles were assembled prior to launch. By October the agency was forced to delay the planned launch of the next shuttle flight by several months, from March to May 2005. These difficulties were compounded when on September 9 the parachutes of the space probe

Genesis failed to open as it returned from a three-year mission gathering solar wind dust, and it crashed ignominiously onto the Utah desert.[29]

The Hubble panel was under its own pressures. At its June hearings, both O'Keefe and Ed Weiler (who in 1998 had been promoted to Associate Administrator for Space Sciences, the position that Homer Newell, John Naugle, Noel Hinners, and Lennard Fisk had all held previously) told the panel that they needed as much information as soon as possible. Weiler had especially emphasized that NASA had to make some important budget decisions by the late fall in order for the robot mission to move forward. Moreover, with the robot mission gearing up, the panel members worried that NASA might do something that would make the shuttle servicing mission option impossible.

By the end of June many members of the panel had serious doubts about the viability of the robot mission. Faber remembered in particular a van ride back from Goddard after seeing a detailed robot demonstration by Cepollina's team. During the ride the panel's robot experts told everyone how unconvinced they were. To them, Cepollina's crew were persistently underestimating the difficulty of what they were trying to do. Rodney Brooks, director of MIT's Computer Science and Artificial Intelligence Laboratory, noted how they didn't seem to understand the required software. Vijay Kumar of the University of Pennsylvania added that the robot mission was made up of many components, all of which seemed plausible until you tried to put them together. Moreover, none had been flight tested, and the robot itself required "one of the most complicated pieces of software ever put together for a robotic system."[30]

In order to put NASA on notice that it had better keep its options open, the panel decided to release an interim letter report in mid-July. In this letter, the panel expressed its strong belief that NASA should not only service the space telescope, it "should take no actions that would preclude a space shuttle servicing mission to the Hubble Space Telescope." Moreover, though the letter report did not specifically oppose the idea of sending robots to service Hubble, it also indicated a concern for the difficulty of the task. "The proposed Hubble robotic mission involves a level of complexity, sophistication, and technology maturity that requires significant development, integration, and demonstration to reach flight readiness."[31]

As the summer ended and the panel began to write its final report, the members increasingly found that the robot mission simply did not make sense. The cost was phenomenal, anywhere from $1 to $1.6 billion and significantly higher than the cost of the single shuttle servicing mission. The time available to make it happen was short, less than three and a half years. And too many of the necessary technologies simply did not yet exist. Though many panel members believed the robot mission could be accomplished, no one believed it could be accomplished in the time available.

On December 8, 2004, the panel released its final report, rejecting almost every decision that Sean O'Keefe had made in connection with the Hubble Space Telescope. On the question of using robots to service robots the Academy panel report was remarkably blunt. "The likelihood of successful development of the [Hubble] robot servicing mission within the baseline 39-month schedule is remote." Instead, the panel recommended that NASA should send the shuttle and humans to service the space telescope. "Meeting the CAIB and NASA requirements . . . for a shuttle servicing mission to [Hubble] is viable," the panel noted, adding that the mission should be flown "as early as possible after return to flight." A robot mission to Hubble should be limited to de-orbiting the telescope, and could wait, since the telescope's orbit was stable and would remain so for at least another decade.[32]

Five days later, Sean O'Keefe announced his resignation from NASA, citing a desire to spend more time with his family.

Timing is everything, however. O'Keefe's resignation so soon after the release of the Academy report suggests that family issues were not his primary reason for leaving. Faced with the Academy report as well as the demands of Congress to change his position, O'Keefe apparently didn't have the stomach to do it. As a "bean counter," he was simply unprepared to deal with the risks entailed with once again flying a space shuttle to Hubble.

Still, O'Keefe agreed to stay on through the release of NASA's next budget in February 2005. Even here, he refused to back down, letting his fear of sending humans into space assert itself in budget negotiations. Having been told by numerous experts that a robot mission to Hubble made no sense, he decided to delete all money for any servicing mission from his 2006 budget proposal, either robot or manned. As

Steve Isakowitz, then NASA's comptroller, explained to reporters, "We have decided that the risks associated with the Hubble servicing at this time don't merit going forward. We don't have the funding to carry out that effort."[33]

It was a scenario that had been repeated many times in the history of the Hubble Space Telescope. In the 1940s and 1950s Lyman Spitzer couldn't get anyone interested in the idea. Then, at the 1962 Iowa summer study he was told the task was too difficult technologically, and should be put off. In the 1970s the idea was at last accepted, but faced repeated attempts to cancel it. In 1974 Congress tried. In 1976 NASA tried. Even during construction in the 1980s the budget problems almost killed the telescope several times. And after launch in 1990 the telescope was almost terminated again when its mirror was found to be warped and it could not focus.

And just as none of these crises had killed the telescope during all those decades, neither did O'Keefe's attempt in February 2005. Only two months after O'Keefe's budget was released, Michael Griffin, his replacement as NASA administrator, announced that he did not agree with O'Keefe, and that he was willing to consider sending the shuttle back to Hubble.

Unlike O'Keefe, Griffin was clear-speaking, concise, and to the point. When someone asked him a question, he gave a precise answer, even if the best he could say was that he didn't yet know. Griffin was also different from almost all of NASA's past administrators in that he had dedicated his life to the exploration of space. He was a licensed engineer and pilot, had taught college courses at three universities on spacecraft design and a variety of aerospace subjects, and had written a textbook on the design of space vehicles. Early in his career he had even worked as an engineer on Hubble's construction. On top of all this he had also worked as an executive for a variety of aerospace firms as well as for NASA. When enlisted for the job of NASA administrator he had been in charge of the space department at Johns Hopkins' Applied Physics Lab in Laurel, Maryland. Unlike any administrator before him, Griffin knew and understood the problems of space exploration. And unlike O'Keefe, he was not afraid to face these problems, squarely and honestly.[34]

During his Senate confirmation hearings on April 12, 2005, Griffin promptly clarified what he thought could and could not be done about Hubble. "I would like to take the robotic mission off the plate," he said,

Fig. 7.4. Michael Griffin during his first official visit to the Kennedy Space Center, May 19, 2005. (Photo courtesy of NASA)

noting how the National Academy panel had called the robot mission "infeasible." He then added, "I believe that the choice comes down to reinstating a shuttle servicing mission or possibly a very simple robotic de-orbiting mission."

Griffin had already put aside the robot de-orbiting mission. There was no rush to launch it, since the telescope was in a stable orbit that would last through at least the year 2020.

Regarding a possible shuttle mission, Griffin was far more sanguine. He explained that O'Keefe's decision to cancel the manned servicing mission "was made in the immediate aftermath of the loss of Columbia. When we return to flight it will be with essentially a new vehicle, which will have a new risk analysis associated with it." He then said, "At that time I think we should reassess the earlier decision in light of what we learn after we return to flight."

In the years that followed, Griffin remained steadfast and firm. As long as the shuttle program moved forward successfully, the shuttle servicing mission to Hubble held a tentative place on the schedule for sometime in early 2008. When the shuttle *Discovery* successfully completed two test missions, proving that the repairs to the external tank worked, and was then followed by a third successful mission of the shuttle *Atlantis*, recommencing construction of the International Space Station, it was time to make a final decision.

On October 31, 2006, Michael Griffin made it official. "We are going to add a shuttle servicing mission to the Hubble Space Telescope to the shuttle's manifest," Griffin announced in his typically careful and precise manner before a packed audience of NASA employees at the Goddard Space Flight Center. "I am fully confident that this fifth [servicing] mission will go as flawlessly as any of us can imagine."

Later that day he added, more emotionally, "This is a day I wanted to get to for the last eighteen months."

Assuming all goes according to plan, in late summer/early fall 2008 a space shuttle, probably *Atlantis*, will blast off from the Kennedy Space Center, once again bringing astronauts and equipment to the Hubble Space Telescope.

The work those astronauts do will hopefully keep the telescope alive for at least another five years. They will replace the gyros and batteries, add new insulation blankets to protect the telescope, and install Wide Field Camera 3 and COS. Moreover, they will try to fix the Space Telescope Imaging Spectrograph (STIS), which had failed in August 2004 when a power supply unit failed, and the Advanced Camera for Surveys (ACS), which had failed in early 2007 when it experienced its own electrical problems. The STIS repair will be especially ambitious, as it will require the spacewalking astronauts to unscrew one hundred and eleven screws to remove an electronics circuit board that no one had ever intended a thickly gloved astronaut to remove.

Finally, the servicing mission will install a soft capture mechanism, essentially a new passive docking port attached to the aft end of Hubble. Hubble had been designed so that only the shuttle's robot arm could grab it, a limitation that had been one of the main obstacles for the robot mission. With a docking port installed, future robot and manned missions will have an easier time attaching themselves to Hubble, thus facilitating the required de-orbit module installation planned for sometime late in the next decade.[35]

Once again, the telescope that has been repeatedly sentenced to death had been given a reprieve.

8

The Lure of the Unknown

Ten months before he announced his decision to service Hubble, Mike Griffin appeared before an assembly of more than two thousand astronomers at the January 2006 meeting of the American Astronomical Society (AAS). Once again, as it had done in 1993 when Hubble was repaired, the AAS was meeting in Washington, DC, and the new NASA administrator was there to give the astronomical community his perspective on NASA's future and its continuing participation in astronomy.

At the time many astronomers were very nervous about their future funding. To some, Bush's new space vision and the resulting shift in focus at NASA from science to exploration seemed a direct threat to their research projects. In an attempt to ease their minds as well as force them to accept the new realities, Griffin was both reassuring and blunt. "It is not our desire to sacrifice present-day scientific efforts for the sake of future benefits to be derived from exploration," he noted. Then he added that, despite this strong support for astronomy, "NASA simply cannot accomplish everything that was on our plate when I took office last April. In space-based astronomy as in other areas we will have to make tough trade-offs."

Before laying out these realities, however, he paused to announce a posthumous award to John Bahcall honoring his service to NASA and astronomy. Bahcall had passed away in the summer of 2005 when a chronic blood disorder had taken a turn for the worse. "In his seventy years John Bahcall was a legend in the astronomical community," Griffin said. "It was John, along with astronomers George Field and Lyman Spitzer, who successfully planted the seed and then tilled the soil for the bold concept that became the Hubble Space Telescope."

To accept the award, Bahcall's wife, Neta, and their daughter, Orli, stepped onto the podium. The room burst into a rousing standing ovation. Then, Griffin discreetly stepped back and let Neta Bahcall make a few comments. She thanked everyone at NASA and the astronomical community for the honor. Then she explained that Bahcall's effort to get Hubble built was based on his basic belief that Hubble would give humanity the chance to discover fundamental new revelations about the universe. She then quoted Bahcall. "'We all have a deep desire to know what exists out there,' John said. 'A desire so basic, so beautiful, so much fun, that it unites all mankind.'" She then smiled and held up the plaque so that her daughter could take her picture.

Once again the room erupted in applause. This time, however, it became a standing ovation that went on and on. Many there either remembered or knew of Bahcall's heartfelt and dedicated effort thirty years earlier to save Hubble from the budget ax and wanted to therefore honor his memory.

The next day at the same AAS meeting there was a press conference to unveil a new Hubble mosaic of the Orion Nebula, with four scientists from universities in both the United States and Canada as well as one scientist from the Space Telescope Science Institute there to explain the image. Because the mosaic had required so many people to produce it from so many institutes, the room was packed with other astronomers. In fact, of the forty or so people in the room, the scientists probably outnumbered the reporters more than four to one.

Before introducing the five panelists, Steve Maran, the AAS press officer and moderator for the press conference, noted that among the crowd of astronomers sitting in the back of the room were several individuals who had played "important roles" in producing the mosaic but could not be fit on the panel. "First of all," Maran said, introducing some of these individuals, "the man who literally wrote the book on the Orion Nebula, the grandfather of the Orion Nebula, Professor Bob O'Dell." Maran was referring to O'Dell's book, *The Orion Nebula: Where Stars Are Born*, written three years earlier for the general public. O'Dell, who had retired from Rice University in 2000 to take a position as a research professor at Vanderbilt University in Nashville, Tennessee, had devoted the bulk of his Hubble research since the telescope's repair to studying the Orion Nebula and the stars being formed within it.

There was polite applause, and O'Dell quipped how the number of people in the room about equaled "the number of people who bought the book." The room broke into laughter, O'Dell sat down, and then the press conference proceeded.

The contrast between these two events was striking. John Bahcall's contribution to Hubble had been prominently noted. Moreover, his sudden death was heartrending, and made people even more willing to stand and cheer his memory.

O'Dell's part in Hubble's creation, on the other hand, seemed strangely forgotten. Both men had worked together and with equal dedication to get the telescope built. Unlike Bahcall, however, O'Dell's lobbying effort had been discreet and off-the-record, since as a government employee it was illegal for him to lobby Congress. In addition, of those who were aware of O'Dell's contribution, too many also remembered his sometimes caustic manner when he was project scientist. Many also could not forget Hubble's misshapen mirror, built under O'Dell's reign. The stain of that error marred their memory of his work on Hubble, and made them less willing to credit him for the achievement of getting Hubble built.

And yet, as much as the telescope had needed John Bahcall's effort, without O'Dell the telescope would not have been built at all. The other instrument scientists had recognized this fact in 1984, soon after O'Dell stepped down as project scientist. At the first science working group meeting following O'Dell's resignation, they held a reception for him in the Goddard library. There they surprised O'Dell by announcing that they had secretly banded together to get him guaranteed time.

"To me, it was a complete surprise," O'Dell said. "Somebody beat on a glass and made this announcement." Knowing that O'Dell had no guaranteed time, each team had agreed to give up a small percentage of its own time; these contributions pooled together would give O'Dell a share comparable to what the interdisciplinary scientists such as Bahcall were getting, equivalent to approximately sixty hours of on-target telescope time. They then presented O'Dell with a picture of the Andromeda Galaxy with etched signatures from all the scientists.

The next day O'Dell called Lyman Spitzer and offered him half of this time, since Spitzer had also been left out in the cold. The two men then collaborated on a spectroscopic program to study the interstellar gas both between the stars and inside the Orion Nebula.[1]

Yet, the contrast in 2006 between how people remembered the contributions to Hubble of Bob O'Dell and John Bahcall remains striking, and epitomizes the many sad ironies of Hubble's story. Though it has paid its builders spectacularly with results far beyond their wildest expectations, the cost for some has sometimes been heavy and painful. Moreover, scattered throughout this story are many unsung or forgotten heroes whose contribution to building Hubble should not go unmarked.

Consider Pete Simmons, for example. Simmons had been an engineer at Grumman Aerospace, working on the early Orbiting Astronomical Observatories. When Grumman took part in the early 1969–70 design studies for Hubble, he immediately fell in love with the idea of a space telescope, and wanted more than anything to help get it launched, built, and used. "It sure would turn me on to have been the program manager of [such] a satellite," he remembered in 1984.

Even before the scientists had launched their lobbying campaign to save the telescope in 1974, he was working to popularize the idea. In 1971 he called the publisher of DC Comics, gave him a tour of Grumman, and convinced him to include the space telescope in an issue of Superman comics. (Superman discovers that the telescope is "saturating the Metropolis area with a strange energy" caused when "the scope passed through a cloud of unpredictable cosmic dust." Superman cleans the telescope's mirror with his heat vision, and all is saved.)

Then, in 1973, before Congress had approved the project and in the days when many astronomers still held doubts about whether a space telescope should be built, Simmons personally conceived and organized an American Institute of Aeronautics and Astronautics (AIAA) conference to expel these doubts. His enthusiasm for the idea was unstoppable. When he first called O'Dell about the idea, he had proposed a symposium held as part of an American Association for the Advancement of Science meeting. O'Dell was unconvinced this would work, and suggested that Simmons hold off.

Only a few weeks later Simmons was back, pushing the idea again with such an infectious enthusiasm that O'Dell couldn't resist. O'Dell suggested that Simmons approach the AIAA. "Pete was the guiding light behind the AIAA topical session on the LST," O'Dell said. "That was really his baby. It wouldn't have happened without him."

Spurred on by Simmons, O'Dell agreed to try to get some of the biggest astronomical doubters of the space telescope to speak, and ended up being remarkably successful. Important astronomers like Allan Sandage, Ivan King, and Gerry Neugebauer agreed to give talks on the use of the space telescope in astronomy. Simmons and O'Dell were even able to convince Jesse Greenstein to appear, something that at first O'Dell had considered impossible. "I think that was a watershed step," remembered O'Dell. The conference made an enormous difference in convincing many fence-sitters in the astronomy community to come down on the side of Hubble.

Simmons's enthusiasm didn't end there, however. During the 1974 and 1976 budget battles his contacts in the aerospace industry were crucial. He not only made his own pitches to representatives and senators, he also acted as a main conduit of information from industry to the scientists and Congress. If Bahcall or O'Dell needed some inside information about the political inclinations of a particular legislator, often it was Simmons to whom they turned to get it. Simmons even cold-called cartoonist Johnny Hart, creator of the comic strip *B.C.*, and convinced him to do a strip on the space telescope's budget problems.

Simmons's support for Hubble was so strong that he was even willing to do the unorthodox. During a 1972 luncheon in New York he was lucky enough to sit at the same table as Governor Nelson Rockefeller. After Simmons had sold Rockefeller on the benefits to New York of the space telescope, Rockefeller wanted to know what he could do to help. "Give me a license plate," Simmons suggested. Rockefeller made the arrangements, and Simmons ended up getting one of the country's first vanity license plates, reading 1-LST.

Without his eager help and determination in the early years when the telescope's future was very doubtful, it is questionable Hubble would have ever been built.

In the end, however, Pete Simmons never helped to build or use Hubble. In 1972, after Grumman lost the space shuttle contract to North American Rockwell, the company's corporate heads decided to get out of the space business. Simmons immediately got a job at Martin Marietta, which was the primary contractor building the Viking Mars landers and which also hoped to get the contract to build the space telescope. However, Martin Marietta had major management and budget problems building Viking. As a result, Simmons began hearing rumors about how the company had no chance of winning the Hubble contract.

Simmons once again went job hunting, and took a position at McDonnell Douglas in 1974, where he worked hard to write their bid for building the telescope. (He worked so hard, in fact, that it almost broke up his marriage.) Unfortunately, McDonnell Douglas didn't get the contract, and when construction started in 1977 he found himself, like Spitzer, locked out of the very project he had helped make possible. Instead, his bosses at McDonnell Douglas gave him work that had nothing to do with space, aviation, or NASA, making him what they called "director of diversification of new products." "I hated it," he later remembered.

In 1981 he got a job working at the newly formed Space Telescope Science Institute working under its first director, Art Code. Code, however, was only there temporarily. When Riccardo Giacconi took over soon after, the two men did not get along and Simmons left, never to work on Hubble again. "It just broke his heart," Bob O'Dell said in 1985. "[He was] one of those guys who just absolutely are caught up with the joy of astronomy and are true believers in the space telescope program."[2]

Nor was Simmons alone. Consider Nancy Roman. During the 1960s and early 1970s there was no one at NASA who was more important in getting the first designs and concepts for Hubble funded and completed. More importantly, it was Roman more than anyone who convinced the astronomical community to get behind space astronomy. "Nancy Roman did a very smart thing when the first round of space telescope proposals came in," noted Ivan King, past president of the AAS and in the 1970s a skeptic of space astronomy. "She called up a lot of good astronomers around the country who hadn't proposed [including King] and got them to be on the panel that looked at the proposals." The result? "She really converted us."

By the time the telescope project finally got started in the late 1970s, however, she was out of the picture. First, her old boss Homer Newell retired. Then, a reorganization at NASA in 1974 placed her position one step lower on the organizational food chain, with a new supervisor who didn't seem to have faith in her. He started giving work to others, including people below her in the chain of command. On top of this, office politics and the bad blood felt by some astronomers due to her hard-nosed approach to funding new projects in the '60s left her isolated. "I somehow became persona non grata," Roman remembered in 1984. Though she did handle much of the negotiations with the European Space Agency concerning their contribution to Hubble, in her last years before retirement in 1980 her involvement with the space telescope was significantly reduced, if not eliminated. Instead, she spent most of her time working on a panel that produced a large but almost entirely ignored report on the future of space exploration.

By the 1990s her work to get the space telescope started in the 1960s was largely forgotten.

On January 10, 1994, Ed Weiler was making the opening remarks at the AAS session that was to unveil the successful repair of Hubble's spherical aberration. In the audience he saw the now retired, sixty-seven-year-old Roman, knitting quietly as was her wont to do at such events. As the man who had replaced her after her retirement, he couldn't let his moment of triumph pass without making sure she received some of the glory as well. He paused, and pointed her out to the audience of 2,000 astronomers. Then he added, "If Lyman Spitzer was the father of the Hubble Space Telescope, then Nancy Roman was its mother." The room burst out into a standing ovation, bringing tears to her eyes.[3]

There were other unsung heroes. The decision to give Westphal's team the contract to build WF/PC didn't just cut Lyman Spitzer out of the loop. In the years up to that decision Spitzer had assembled a cadre of top-notch engineers and scientists at Princeton to design and build the camera, including engineer Jon Lowrance and researcher John Rogerson. For almost a decade these men labored to perfect the SEC Vidicon detector. Though their work showed that a space telescope could be built and therefore helped get it built, when the contract went to JPL they, like Simmons and Spitzer, were left out in the cold.

Then there is the story of astronomer Bob Danielson. Danielson had originally been hired by Martin Schwarzschild to work on the Stratoscope balloon telescopes, and in later years ended up unofficially leading the project under Schwarzschild. Later, as a professor at Princeton, he became an original member of the science working group that did the initial design work on Hubble in the early 1970s. "He was very, very key to the formative phases of the space telescope," Jon Lowrance remembered. His experience building remotely operated telescopes such as Stratoscope, combined with his passionate research into galaxies and planetary atmospheres, made him the ideal person to formulate the specifications for Hubble's cameras. As Schwarzschild noted, "Bob Danielson showed himself, both as a scientist and instrumentalist and as a person, magnificent."

Even as he was doing this important work, Danielson did so knowing he was dying of leukemia. When he passed away in April 1976 after a long battle, it sent shock waves through the Hubble community. "It was one of the worst days of my life," remembered John Bahcall in 1984. "I loved him. He was courageous, magnanimous, very intelligent, wonderful companion, a lot of fun." Some believed that it was the loss of Danielson's influence that contributed to Princeton losing the camera as well as the institute contracts. "I think Princeton's role in the space telescope wouldn't have gone down as much as it did, because he would have been more involved," noted Lowrance.[4]

Then there is Jean Olivier, who spent almost his entire career working as Hubble's chief engineer. As a kid Olivier liked to tinker with cars, and so it was a natural for him to get a degree in mechanical engineering and go to work for Chrysler in Detroit, where he earned a master's in automotive engineering from the Chrysler Institute of Engineering. At the time, 1958, however, jobs designing cars were scarce, while jobs building rockets were booming. Chrysler was doing support work in Alabama building rockets for von Braun. Olivier went there, eventually switching from Chrysler to NASA when the chance came. By 1969 he was in charge of the astronomy office at Marshall, where he supervised Hubble's early design work. When in 1973 Bob O'Dell took his "dog and pony show" across the country and Europe to sell the telescope to astronomers, Olivier went with him to make the engineering case. Then, in 1974 he was named Hubble's chief engineer, a position he held through every crisis until 1988, when he became deputy manager of the Chandra X-ray telescope project. He returned to Hubble, however, during and after launch in 1990, acting as chief engineer for the orbital checkout team. "[He] has been on the program since Day 1," Jim Odom said soon after he took over as project manager from Fred

Speer. "[Olivier] is my technical conscience relative to the performance of the hardware."

Yet, though he more than anyone had been in charge of building the telescope, once it went into operation his part was over. For one thing, the deal had always been that operations would switch from Marshall to Goddard after orbital checkout. For another, the discovery of spherical aberration marred his credibility, just as it had O'Dell's.[5]

Then there were those who have not only been largely forgotten but were forced to fall on the sword for the sake of the telescope. Consider Fred Speer. When he took over as manager of Hubble in 1980, he immediately recognized that the telescope was over budget and that people were being unrealistic about its costs. Yet, he also knew that if he demanded a realistic budget from his bosses, Marshall director William Lucas or NASA administrator James Beggs, they would refuse. Caught between a rock and a hard place, he improvised, cutting the telescope itself to the bone. Even this wasn't enough, and when things finally fell apart in late 1982 and early 1983, he ended up the fall guy, removed as manager so that those who had actually caused the problem could claim credit for fixing it.

Even among the individuals whose contribution we remember, the cost has sometimes been heavy. Like O'Dell's, Steve Beckwith's campaign to save Hubble did not help his career. On July 16, 2004, he announced that he would resign his directorship of the institute when his term expired in September 2005. Though he could have stayed, the institute's contract with NASA was soon to expire, and his aggressive advocacy for Hubble had not made him well liked by some at NASA headquarters. Rather than remain and become a distraction that might even risk the institute's contract, he chose to move on, letting others take up the torch.

And then there is Robert O'Dell. Despite the gift of guaranteed time as well as the later award of more Hubble time in straight competition, O'Dell really got very little personal reward for his ten-plus years of effort to get Hubble built. And it seems somehow wrong that his participation in this achievement, like Olivier's, has been marred by the spherical aberration. Like Speer, O'Dell has had to carry the blame for problems caused by those above him.

One of the primary reasons that Perkin-Elmer had been chosen as the contractor to build the mirror assembly was its successful experience building similar mirrors for the military. O'Dell, who had had a top secret clearance since the 1960s, was aware of P-E's work, and had been impressed by it. With the military, however, the budget arrangements were far different. The military had plenty of money. It also did not like to micromanage its contractors. Instead, it would tell them what was needed and let them build it. If the contractors ran into problems the military would simply give them more money. This was how P-E expected to work when it began building Hubble.

Unfortunately, this was not how NASA could function, especially during the late 1970s and early 1980s with a reluctant Congress and limited budgets. First, because of P-E's military work, the space agency was forbidden to assign its normal contingent of managers to the work, thereby denying it the ability to manage—no less micromanage—P-E's operation. Second, and more important, NASA simply didn't have any extra cash. James Fletcher had decided that the telescope would cost $300 million in 1972 dollars, and that was that. As a result P-E simply didn't have the resources to do the kind of testing and oversight it normally would, and was under such budgetary and time pressures that its managers and engineers felt inclined to look the other way when something unusual occurred, such as the need to add washers to the reflective null corrector. It would have cost them a lot of time and money to check and correct this issue, time and money they simple didn't have and NASA wasn't going to give them.

We should therefore not be too terribly surprised that the primary mirror was ground incorrectly. The pressure to meet Fletcher's impossible budget goals made the occurrence of an error like the spherical aberration almost guaranteed. In fact, in retrospect it is both a miracle and a testament to the dedication of everyone who worked on it that Hubble wasn't launched with many more problems.

After O'Dell stepped down as project scientist, he made one last effort to gain some personal reward for his years on Hubble. At the time, the shuttle program was at its height and NASA had very flexible rules for selecting shuttle astronauts. O'Dell, whose passion for flying made him an ideal candidate among the astronomers to become an astronaut, campaigned to get himself named as a payload specialist to fly on the shuttle mission that would deploy Hubble.

Though the other scientists in the working group were generally not enthused, O'Dell kept pushing the idea. In August 1985, with Hubble's launch then scheduled for about a year later, O'Dell was in Austria competing in the world flying aerobatics competition when he got a call from Jim Odom, telling him he had been selected to fly in space. "My little feet were hardly touching the ground," was how O'Dell described to me his joy at that moment. Then he joked, "I was probably unbearable."

Unfortunately, the *Challenger* accident in 1986 changed everything, squelching O'Dell's only chance to fly in space. For O'Dell, the big payoff for years of work on Hubble had once again fizzled.[6]

In fact, the number of dedicated people who worked many years to either build or save Hubble without fame or credit or to their own personal detriment—such as O'Dell, Simmons, Roman, Danielson, and Speer—is countless, and unfortunately they cannot be listed completely here.

Even more amazing are the many people who originally opposed the telescope (like Jesse Greenstein and Riccardo Giacconi) or didn't care whether it was built or not (like James Fletcher) who found in the end that they could not resist its allure, and ended up providing help and aid in getting it built. Even Sean O'Keefe, who couldn't bring himself to launch a manned mission to Hubble, also couldn't let the telescope simply die, and had been willing to spend billions that NASA didn't have for a robot servicing mission instead.

In the end, the undeniable appeal of Hubble, the compelling nature of the unknown that it promised to unveil, was what finally won out, beating back every form of opposition or skepticism from untold naysayers while simultaneously compelling its supporters to make sacrifices surprising even to themselves.

On a basic scientific level, the need for a good optical telescope to complement the telescopes working in gamma-, x-, infrared, ultraviolet, and radio wavelengths simply made sense, for astronomy itself. Or as Spitzer himself explained in 1977:

> In the past at any rate, astronomers have developed the concept of a balanced program, in which optical space astronomy, ground based

> astronomy, infrared, x-ray, radio astronomy, all move forward together at the maximum pace that the total funding will permit. . . . You really need research and results in all these fields, to explore the universe in the most effective way.[7]

On a more fundamental human level, however, Hubble epitomized the inevitable and irresistible lure of the unknown. Spitzer understood that attraction in 1946 when he said that a space telescope would "uncover new phenomena not yet imagined." Bahcall repeated the point years later when he said, "We all have a deep desire to know what exists out there." From Henry Norris Russell to Ernst Stuhlinger to Pete Simmons to Bob O'Dell, everyone knew that a telescope above the atmosphere was going to see things that had never been seen before. No one could resist that promise.

Nor were the scientists and engineers the only ones captivated by this lure of the unknown. It is present in all of us. Consider for example a story told by one of Hubble's builders in 1984. Evan Richards was then project manager for the High Speed Photometer being built at the University of Wisconsin. At the same time the university had a program called College for Kids. Grade school children were brought to the university to expose them to the world of research and intellectual creativity. As Richards explained in 1984, "It ranges all over the map, from radio and TV stuff to drama, music, chemistry, whatever."

In 1983 Richards was asked to give a presentation about astronomy to one grade school group. Before the kids were brought into the room Richards wrote two numbers on the blackboard: 1609 and 1985. The first was when Galileo made his pioneering telescope observations. The second was Hubble's then expected launch date.

The kids were brought into the room by three elementary school teachers, who introduced Richards. "They didn't mention space telescope," Richards remembered. Instead the teachers were vague, saying that he worked on space projects, engineering and science, and was "going to tell you some things."

Richards started by pointing on the blackboard and saying, "I wrote two dates on the blackboard there, 1609, 1985. Does anybody know why I did that?"

Instantly about half the kids raised their hands. Richards picked one at random, who immediately said, "Well, 1609 was when Galileo first made some observations with the telescope. 1985 is when you expect to launch the space telescope."

"The teachers, their jaws dropped," Richards remembered. "They changed color visibly." No one had primed the kids. They had not even been told what Richards was going to talk about. Moreover, the boy who spoke up was not someone the teachers had even expected to raise his hand.

Yet, the boy had not just spoken up, he had understood the significance of Hubble's launch. "What was always interesting to me," remembered Richards, "was how well informed grade school students were on some subjects. Sometimes they amazed their teachers with their questions and answers."

This kind of random event occurred repeatedly to Richards as well as many other people involved with building Hubble, and included both kids and adults inside the United States as well as everywhere across the globe. Hubble had an intrinsic potential for uncovering the unknown, and almost instinctually the human race had been drawn to it.

We should therefore not be surprised at the project's inextinguishable history. From the beginning, Hubble represented a basic human urge that could not be denied.[8]

■ ■ ■

Despite Hubble's phenomenal success, the unknown still lurks out there in space, requiring a variety of new telescopes in all wavelengths. In optical wavelengths, the immediate future appears somewhat brighter because of the decision by Michael Griffin to reinstate the shuttle servicing mission to Hubble. Assuming the 2008 shuttle mission succeeds, we can look forward to at least five more years of extraordinary discoveries from Hubble.

The long-term future of optical space astronomy, however, remains problematic. Even if the servicing mission successfully upgrades Hubble, the most that mission can accomplish is to extend the telescope's lifespan until sometime into the middle of the next decade. By that time—if nothing else is done—the telescope's new gyroscopes will begin to fail and the world's only optical space telescope will reach the end of its life.

Hubble's limited life expectancy is even more significant because, of the more than two dozen space telescopes under construction or design at this time, not one is a general optical observatory like Hubble. Of the handful that will work in optical wavelengths, all have very specific or limited research goals. For example, Kepler, launched in March 2009, is designed only to search for extrasolar planets. For four years it will be aimed at one specific area on the sky, looking for the transits of planets across the face of the stars in its view, evidenced by a dip in light. Similarly, when Terrestrial Planet Finder is launched sometime in the middle of the next decade, its primary goal will be to detect Earthlike planets orbiting approximately 150 stars within a 45-light-year radius of the Earth. Neither of these optical telescopes is intended to provide astronomers a general tool for doing astronomical research.[9]

And though NASA officials and astronomers often describe the James Webb Space Telescope (once called the Next Generation Space Telescope) as a Hubble replacement, such statements are not quite accurate. Though Webb will be a powerful successor to Hubble, it is not an optical telescope. It is being built to operate in infrared wavelengths, as recommended by the astronomy panels in the 1990s. Moreover, though Webb is being planned as a general observatory like Hubble, its primary focus is to do cosmological research, following up on the Hubble deep field images by looking at the most distant events taking place only a few hundred thousand years after the Big Bang.

Nor will ground-based adaptive optics be able to fill the gap left when Hubble goes down. Though some astronomers still like to tout the ability of adaptive optics to compensate for the atmosphere's fuzziness, noting how advances have been made in using lasers to create a simulated target star to calibrate the atmosphere's fluctuations, such technology still only works in the near-infrared range, and its images can only cover a tiny field of view. As noted in the final report of the 2004 National Academy of Science panel on Hubble's status, "Ground-based adaptive optics systems will not achieve Hubble's high degree of image stability or angular resolution at visible wavelengths for the foreseeable future."[10]

Thus, when Hubble finally fails, researchers who study star formation and evolution, planetary nebulae, extrasolar planets, molecular clouds, novae and supernovae, and a whole host of other relatively nearby Milky Way phenomena will lose a invaluable tool. Though they still will be able to do wonderful and significant research in other wavelengths, such as in the infrared with Webb, without the availability of sharp optical images as well as ultraviolet spectroscopy scientists will struggle to interpret their data in these areas of research.

And without optical images like Hubble's to give clarity to most astronomical research there will also be far less general interest in the subject. As much as astronomers like to think that the general public loves astronomy, the truth is that what ordinary people really care about are things that they can see, with their own eyes. As one ham radio enthusiast and taxpayer once told me in discussing Hubble, "I just want to see the pictures."[11] Without the availability of spectacular optical images like Hubble's, it will be much more difficult to excite children, such as the ones Evan Richards taught in 1983, about the subject of astronomy.

Faced with this impending deadline, the astronomical community does have several options. One very unlikely choice is for NASA to take the advice of John Bahcall's August 2003 committee and send an additional shuttle servicing mission to Hubble around 2010, just prior to the shuttle's retirement. This is the kind of work for which the shuttle was devised, and for which no space vehicle presently under design or construction will be capable of matching.

Unfortunately, present-day politics makes this choice improbable. In the political arena everyone from both political parties has apparently accepted President Bush's decision to retire the shuttle fleet in 2010. Until then the shuttle schedule is full, with little margin. And to send another shuttle to Hubble in 2010, only two years after the planned servicing mission, seems extravagant.

Another option is to upgrade the eventual Hubble de-orbit mission. No matter what else happens, sometime in the next decade NASA will have to send another mission to Hubble to attach a de-orbit engine. Otherwise when the telescope's orbit finally decays there will be no way to control its re-entry.

That de-orbit, however, doesn't have to happen immediately. Instead, the mission—at relatively little cost—could be expanded to give Hubble another set of new gyros and new batteries, even new instruments, and extend the telescope's life further. Moreover, this mission could be done either by a manned or robot mission. If by robot it could use some of the same technologies that Frank Cepollina's team at Goddard developed in 2004, this time with ample time to test and work out the kinks. If manned it could use the Orion space vehicle that NASA is now building to make possible the human exploration of the solar system.

All of these plans have one inherent weakness: they assume that nothing else will go wrong with Hubble in the interim, an unlikely possibility considering that the telescope has already completed its original fifteen-year mission and by 2015 it will have been in orbit for a quarter of a century. The loss for example of Hubble's Advanced Camera for Surveys in early 2007 because of a failure in its power system was clear proof of this reality.

The only real long-term solution for maintaining our ability to observe the heavens clearly in all wavelengths is to begin the challenging work of replacing Hubble. Such a replacement telescope need not be another gigantic government project like Hubble. Many universities today run small programs designing and launching student-built Earth-observation mini-satellites. There is even a university project, funded by the Education Department of the European Space Agency, to build and launch a student-built lunar orbiter. Such programs, teamed with amateur and professional astronomers as well as the private sector, could just as easily produce a plethora of small, low-cost optical space telescopes able to do significant research, independent of government funding. A telescope manufacturer could even donate a small commercially made telescope to an astronaut for use on either the Moon or the International Space Station, giving it the proper programming so that it can track objects in the sky. This would be a simple and cheap way to allow some astronomical research to continue in space, and the publicity would be priceless.

In another option, some astronomers have improvised their own space-based telescopes, using instruments designed for other purposes and co-opting them to become general observatories. The most prominent example is the Swift Gamma Ray Telescope, launched November 20, 2004, and designed to study gamma ray bursts. Carrying three separate instruments—a gamma ray detector, an x-ray telescope, and an ultraviolet/optical telescope—Swift normally monitors the sky for gamma ray bursts, immediately slewing to one when it occurs so the x-ray and ultraviolet/optical telescopes can get images of the burst within seconds. Since it only detects a burst about once or twice a week, however, it has lots of dead time, during which it is used to observe a variety of astronomical targets, from comets to supernovae.

Similarly, the Galaxy Evolution Explorer (or Galex, as it is known among astronomers) is an ultraviolet telescope designed to study the evolution of galaxies across 80 percent of the sky. Launched on April 28, 2003, it has also been used by astronomers to study a host of other subjects, from globular clusters to white dwarf stars.

The abilities of these two telescopes—as well as the smaller telescopes that universities can build—are limited, however. The mirrors of Swift and Galex are less than 12 and 20 inches across respectively, far less than Hubble's 94-inch diameter. None of these small telescopes can replace Hubble. In order to match or supersede Hubble's optical/ultraviolet abilities, you need to launch an actual Hubble replacement. And to get it built before Hubble fails, design work must begin immediately, since such government projects usually take between ten and twenty years from conception to launch, and Hubble is unlikely to last that long, no matter what we do.

In this context we do have some hope. With the advent of NASA's new program to return to the Moon the astronomical community has been offered a wealth of opportunities and tools for getting new telescopes into space. As part of this lunar program NASA is building two new rockets from shuttle-derived technology. Ares 1 will be equivalent to the Saturn 1B rocket used to launch crews to *Skylab* in the early 1970s. Based on the shuttle's solid rocket boosters, it will be the crew carrier of the system, putting the Orion capsule into Earth orbit.

The Ares 5 in turn will be akin to the Saturn 5, though far more powerful. In fact, it will be the most powerful rocket ever built, able to put as much as 130 tons of usable cargo into orbit, thirty percent more than the Saturn 5 and more than four times as much as the shuttle. Used as a cargo carrier, it will bring crews the additional modules and supplies they will need to travel to the Moon and beyond.

Ares 5's size will also make it an excellent launch vehicle for future large telescopes. In fact, this rocket will put so much mass into orbit that there will always be room to launch a variety of scientific experiments, including those focused on astronomy, and send them anywhere in the solar system.

With this in mind astronomers have already begun proposing experiments for placement on the Moon when NASA establishes its bases there. For example, the Moon's far side is probably the best place in the solar system to do radio astronomy, since the plethora of television, radio, and telephone broadcast signals from Earth are blocked from reaching it. Since NASA intends to build its first outposts at the lunar poles, sending astronauts the short distance to the far side is not unreasonable. Using some remarkably lightweight and simple radio telescope designs—in one case nothing more than a rolled up piece of Mylar several hundred feet long with the metal radio antenna imprinted on it at regular intervals—astronauts can quickly lay out a radio telescope able to do cosmological research.

An even simpler idea might allow astronomers to solve the mystery of the universe's accelerating expansion rate. By installing small reflective mirrors on the lunar surface and then aiming Earth-based lasers at them, scientists can make extremely precise measurements of the distance between the Earth and the Moon. Based on some theories, this information could explain the universe's expansion rate.

Putting an optical telescope on the Moon is of course more complicated, and will obviously have to wait until the human bases on the lunar surface are better established. However, several astronomers have already begun thinking about this problem, proposing that any lunar telescope facility be treated not as a mission with a limited lifespan—like Hubble—but as a ground-based facility, begun small with only a few instruments, but once established, allowed to grow with time—much as Bob O'Dell proposed for Hubble on Black Saturday back in 1980.

Even so, for optical astronomy the Moon has its limitations and problems. Putting a telescope on the lunar surface entails extra cost. On the ground you can only see half the sky at any point during the year, and depending on where you are some areas on the sky will not be visible at all. Furthermore, the lunar surface is covered with a fine-grained angular dust, which can foul equipment and telescope mirrors. Worse, there is even some evidence, still unconfirmed, that this dust can get elevated by electrostatic forces, thereby producing an extremely thin atmosphere that nonetheless could subtly obscure the sky.

In the post-Hubble era, however, humans have on hand a wealth of experience operating free-flying orbiting telescopes, a skill no one wants to lose. Rather than put the biggest future telescopes on the Moon or in low Earth orbit (as Hubble was), most astronomers are now aiming to put their instruments at L2, a point in space where the gravitational forces between the Earth and Sun balance. L2 is located about a million miles away from the Earth, on its night side away from the Sun. Here the Wilkinson Microwave Anisotropy Probe (WMAP) was placed in 2003, where it measured the sky's cosmic microwave background. Similarly, the James Webb Space Telescope will be placed there when it is launched in 2013. At this location, a telescope will not require much fuel for holding its position, and it will have a continuous view of the entire sky.

The disadvantage of L2, however, is that it is far away from the Earth. Sending a crew there to do servicing, as we have done with Hubble, will not be practical for many decades to come.

However, there is a solution. Astronauts on their way to the Moon automatically pass through the L1 point of the Earth-Moon system, a point where the Earth and Moon's gravities balance. The amount of energy required to travel between this L1 point and the Earth-Sun L2 point is not exorbitant. If future telescopes at L2 are given the engine capability to return to L1, they will be accessible to Orion crews on their way to and from the Moon. Give the Orion system its own capability to do servicing, including a robot arm, cargo capacity for instruments, and the right kind of spacesuits and tools, and it will be possible for humans and telescopes to rendezvous, station-keep at L1 for a few days while the astronauts do repairs and upgrades, and then go their separate ways.*

In this context the James Webb Telescope is already being outfitted with a simple grappling fixture, on the off chance a future servicing mission, not yet conceived, will want to grab and fix it. Moreover, as NASA engineers are right now doing Orion's first design studies they are considering incorporating some of these servicing capabilities into the system.

At first glance getting these proposals built might seem far-fetched. Yet, all signs, both political and cultural, appear to favor their achievement. On the cultural side, the astronomical community today has embraced manned space exploration in a way that is unprecedented. Unlike in the 1960s and 1970s, most astronomers are no longer arguing that the

* As of the spring of 2007 it is still unclear whether Orion will have these servicing and spacewalk capabilities. Some design concepts being considered by NASA actually exclude all spacewalk capabilities from Orion.

money spent on manned space exploration would be more effectively spent on unmanned telescopes. Instead, in another example of Hubble's significant cultural influence, their experience of having Hubble fixed, maintained, and upgraded by astronauts has convinced them that any future telescope should follow the same paradigm. As noted in 2006 by Matt Mountain, the director of the Space Telescope Science Institute, "None of [Hubble's discoveries] would have been possible without a partnership between the human spaceflight program and science." Astronomers want their future unmanned observatories serviceable and repairable by humans, just like Hubble.

There are still issues that the astronomical community has not yet dealt with. Almost every space mission that astronomers are proposing and NASA is funding are like Kepler, Swift, and Webb, with more specific scientific goals than Hubble. Though their use can and will be expanded once in space, as has been done with Swift, and their goals certainly do focus on some of the most critical scientific questions of the day, their very specificity in design naturally limits their capabilities as general observatories.

None of these issues are real deal breakers. As the American space program moves forward to the Moon, it appears that astronomers want to move forward with it.[12]

Politically, the situation for NASA and the American space program has probably not been this rosy since the early 1960s. Consider, for example, senator John F. Kerry (D-Massachusetts), whose space platform during the 2004 presidential campaign was hardly as enthusiastic as Bush's for the human exploration of the solar system. Nevertheless, Kerry included manned exploration in his overall platform, suggesting the U.S. space program must balance human exploration with the need to do astronomical, planetary, and aeronautical research. As stated by Lori Garver, a member of Kerry's Science and Technology advisory team, "We will support solar system exploration as an important goal for our human and robotic programs . . . but only as one goal among several."[13]

Nor was Kerry's campaign position unusual. On July 16, 2006, Barbara Mikulski (D-Maryland) and Kay Bailey Hutchinson (R-Texas) introduced a bipartisan amendment to add one billion dollars to NASA's budget in order to pay for the costs for getting the shuttle redesigned after the *Columbia* accident. Though their proposal was not adopted, that two of the most powerful elected officials in the Senate—from both parties—were pushing for an increase in NASA's budget indicates the general support for space exploration that exists nationwide.

And though many people—from academics to politicians to aerospace experts—still strongly disagree with Bush's specific proposal to return to the Moon, few have adopted the position that human space exploration is unnecessary, as many had in the past. Instead, they simply have argued that the nation must do it differently. Though administrations can change and with such change policy can shift, the signs all indicate that the United States is firmly committed to making the human settlement of the solar system happen, one way or the other.

The significance of these political and cultural trends for the American space program is truly profound. After more than forty years of debate, it appears that the argument about whether to fly manned or unmanned missions in space is over and the supporters of manned spaceflight have won. It has become accepted—mostly due to Hubble—that the best way to proceed with the scientific exploration of space is for humans and robots to do it together.

Reinforcing these positive signs has been what seems the shrewd political judgment of Michael Griffin. Compare his actions from 2004 to 2008 with those of James Fletcher from 1972. Griffin has been tasked by the President and Congress to finish the International Space Station using the space shuttle, service Hubble, build a new manned spacecraft, retire the shuttle, and begin lunar exploration, all by 2015.

Yet, as in the 1970s, Congress and the President have been reluctant to give him sufficient funds to achieve these ambitious goals. When Bush announced his new "vision" in 2003 he only offered NASA one billion dollars extra, and insisted that this extra capital be spread over a five-year period. In turn, Congress ate into this limited budget in 2004 by forcing upon NASA a total of a half billion dollars of unexpected earmarks. Somehow, NASA was expected to meet all its goals as well as build a plethora of pork barrel projects that had nothing to do with returning to the Moon.

Unlike Fletcher—as well as almost every other past NASA administrator—Griffin has not made believe that he can do all this for the money given. Instead, he has bluntly told Congress that if he doesn't get the funds he will simply not do it all. In this context we shouldn't be surprised that Mikulski and Hutchinson offered to boost NASA's budget. For the first time in decades, they have a NASA administrator whose word they can truly trust.

Thus, if the astronomical community pushes to have an optical telescope built to replace Hubble, and Griffin decides to push for it, his creditability with Congress enhances his chances of convincing them to pay for it.

And if this particular optimistic view of the future fails to unfold? No matter. The outcry to keep the Hubble Space Telescope flying in the last few years has illuminated one fact that too many in the past have been unwilling to face—that before Hubble the human race could not see the sky as it really was, and after Hubble no one wants to return to that state.

■ ■ ■

On March 31, 1997, Lyman Spitzer passed away at the age of eighty-two. Despite the heartaches and defeats he had experienced in getting the Hubble Space Telescope launched, things had not turned out that badly for him in the end. Not only did O'Dell's gift of guaranteed time allow him to do research with Hubble, he had been appointed to and chaired several advisory committees set up by AURA, including the Space Telescope Institute Council, which provided oversight and advice to the Space Telescope Science Institute in order to help maximize the telescope's scientific output. In addition, Spitzer served as Princeton's representative to AURA's board. These positions meant that during the last years of Hubble's construction Spitzer was intimately involved in making sure the astronomical community—as well as the public—got the most out of Hubble. "From my own personal standpoint it's worked out better," Spitzer said in 1983. "I have a much more important position at AURA and the [institute] than I would have had if the science institute had been at Princeton."

Thus, though he never really got the chance to use the telescope as he wanted, he ended up making sure everyone else could. For this every astronomer I have ever spoken to about Spitzer is eternally grateful.

And though the loss of the camera and institute were heartbreaking, Spitzer did have his moment of glory in space, though not with Hubble.

He had been the principal investigator for the Copernicus ultraviolet space telescope, launched in 1972 as the last of NASA's Orbiting Astronomical Observatories.

For him, the pressure of the situation was excruciating. At the time of launch Spitzer had been one of the country's leading figures pushing for the construction of space telescopes for almost a quarter of a century. Yet, two of the previous three OAO telescopes had failed, the first at launch and the second after only a few hours in orbit. Now, *his* space telescope was about to go into space.

The day before launch he was down at Cape Canaveral, reviewing his notes and trying to make sure everything was as it should be. To his horror he suddenly realized that the telescope's secondary mirror had been set at the wrong focus position. Worse, the problem was embarrassingly related to "a mistake in converting from centimeters to inches." He quickly began calling engineers and mission controllers, getting them out of dinner to go down to the control room to fix the problem. "The last thing we did before they cut off communications [prior to launch] was to command the focus motor to move the secondary mirror to an appreciably different position. Immediately afterwards, the umbilical cord was disconnected and we were cut off."

Once in orbit, they discovered that the focus drive on the secondary had failed, locking the mirror in this last focus position. Fortunately, it was the right position, so Spitzer's first space telescope was in focus from the start, unlike Hubble.

Still, for Spitzer the real moment of truth came four days after launch. Because of the high voltages of the scientific equipment, the instruments were initially left off so that the satellite could vent any remaining gases and thus avoid any voltage arcs that could fry everything.

After four days, it was time to turn everything on. For this event Spitzer came down from Princeton to the control center at the Goddard Space Flight Center in Maryland. He was too nervous to sit around, however, and found himself walking the center's grounds. "I was preparing myself for the worst," he explained. "'Be prepared to accept whatever happens,' I told myself."

When the instruments were activated, however, they worked perfectly. In fact, Copernicus worked so well that it functioned for over nine years, far longer than originally planned, making ultraviolet observations of thousands of objects. For Lyman Spitzer it was a true triumph, proof that his 1940s concept of an orbiting space telescope was valid.

Ironically, Lyman Spitzer's life in science ended up doing as much good for the sport of mountain climbing as it did for astronomy. In 1985 Spitzer had been awarded the Crafoord Prize from the Royal Swedish Academy. Equivalent to a Nobel Prize for astronomy, the prize included an award of $135,000.

Soon after, he and his climbing partner Ted Cox were driving to the Shawangunk Mountains in upstate New York for another day of rock climbing. The Gunks, as climbers call them, are one of the best climbing spots in the eastern United States, and are thus extremely popular. As they drove, Spitzer told Cox about his award, and then mentioned that he really didn't know quite what to do with the money.

"Well, after you finish giving me my share," Cox joked, "I think you should give the rest of it to the Gunks."

Spitzer got very quiet for a long time. Then he said, "That's not such a bad idea."

Spitzer subsequently changed his will to include endowments of one million dollars each to the Mohonk Preserve, which owned the Shawangunk Mountain climbing area, and the American Alpine Club. At the Gunks his gift allowed the Mohonk Preserve to install safety bolts at some of the more popular climbing routes. And his gift to the American Alpine Club established the Lyman Spitzer Climbing Grants program, awarding grants to "small, lightweight climbing teams attempting bold first ascents or difficult repeats of the most challenging routes in the world's great mountain ranges."

In the final analysis, none of us can ever truly predict the consequences of our actions. Lyman Spitzer's passion for building a space telescope so that humanity could see the universe's unknown mysteries ended up also making it possible for rock climbers the world over to climb mountains they never would have dreamed possible.

And if that isn't a wonderful metaphor for everything that the Hubble Space Telescope represents, I don't know what is.[14]

Fig. 8.1. Lyman Spitzer on Baffin Island. (Photo by Don Morton)

AFTERWORD TO THE PAPERBACK EDITION

Like any book about the events of its time, *The Universe in a Mirror* was quickly overtaken by history since its publication in hardcover in 2008.

First, the long-delayed shuttle servicing did not occur in October 2008 as scheduled. Instead, it was scrubbed only two weeks before launch, when Hubble's Science Instrument Control and Data Handling Unit failed. It is this unit that transmits almost all of the telescope's scientific data, including images, to the ground. Without it Hubble was useless.

Fortunately, the unit had been designed with redundancy, a primary A side and a backup B side. Engineers at the Goddard Space Flight Center were able to switch operations from the failed A side to the B side with few problems, and within weeks the telescope was once again transmitting data back to Earth. However, without a working A side the unit no longer had redundancy. Should the B side also fail, there would no longer be any way to collect Hubble's scientific data.

The data handling unit, however, was also an "orbital replacement unit," designed for easy astronaut removal and installation. Moreover, a working spare already existed, stored at Goddard and used as a test bed for Hubble's new instruments. For astronauts to swap out the old unit with this spare would be a relatively trivial task.

NASA officials therefore decided it was foolish to fly the servicing mission and not make this swap. However, to get the spare unit checked out and ready, as well as train the astronauts in its installation, required several months of preparation. Thus, the mission was delayed seven months to May 11, 2009.

When May finally arrived and the shuttle *Atlantis* took off, the actual mission went off like clockwork. Not only did the astronauts replace the data handling unit, they achieved all the other planned repairs and upgrades and did the work so efficiently that they almost made it look easy. Even the repair of the Space Telescope Imaging Spectrograph (STIS), requiring the unscrewing of 111 tiny screws to remove a failed circuit board, went smoothly.

The only serious glitches during the mission involved a bolt that would not unscrew and a screw that became stripped. In the first case, the bolt needed to come free in order to remove the old Wide Field/ Planetary Camera 2 (WF/PC-2), which had been installed in 1993 during the first servicing mission to repair Hubble's spherical aberration. The bolt was on too tight, and would not unscrew when astronaut Andrew Feustel tried to remove it using his power screwdriver in automatic mode. Finally, Feustel was given the go-ahead to try to unscrew the bolt manually. If the bolt became stripped or broke off, it would mean that the sixteen-year-old WF/PC-2 would remain on board and that the new Wide Field Camera 3 would have to be returned to Earth, unused.

The manual effort worked. "Woo hoo," exalted Feustel. "It's moving out!"[1] The new camera was then installed without a hitch.

Then there was the stripped screw. To fix STIS, astronauts Mike Massimino and Mike Good had to remove a cover plate held on by 111 screws so that they could then replace a failed circuit board. Removing the cover plate, however, required that they first remove a handrail attached to the telescope and held on by four screws.

The first three unscrewed easily. The fourth, on the bottom right, however, would not turn, and in his effort to get it free with his power screwdriver, Massimino stripped the screw's head. After some quick experimentation on the ground, Massimino was told to grab the handrail and literally rip it off manually. Ground calculations indicated that if he pulled the top of the handrail outward, the screw at the bottom would simply snap, freeing the rail.

The calculations were right. "I can feel it now, starting to come away," Massimino said as he applied force. "Here we go!"[2] With the handrail out of the way, the subsequent removal of 111 screws and the installation of a new circuit board on STIS were completed so smoothly that it almost seemed boring.

That these repairs, as well as all the work on Hubble, went so well is especially ironic in that doing this kind of work in space is *not* easy. By making it seem trivial, the astronauts once again helped convince ordinary Americans that space exploration is routine and straightforward, something that it will not be for centuries to come.

Regardless, the remarkable success of the shuttle servicing mission means that Hubble now has a full set of new batteries, new gyroscopes, two working cameras, and two working spectrographs. Unless something unexpected goes wrong, the telescope should remain operational for another five years at least.

The shuttle servicing mission was not the only noteworthy event to occur since the book's publication in 2008. There have also been significant changes in the political landscape, the election of Barack Obama in November 2008 being the most significant. Whether his administration will continue George Bush's proposal to return to the Moon remains unclear at the time of this writing (fall 2009). At a minimum the Obama administration has been reviewing the plans to build the Ares 1 and Ares 5 rockets, and is even considering canceling their construction. How such an action would affect the field of astronomy is also unclear, since many astronomers have linked their future space telescope proposals, including a new proposal for an optical space telescope, on the heavy-lift Ares 5 rocket. If that rocket is not built, such plans would have to be scaled back significantly.

The arrival of a new president also meant the end to Mike Griffin's regime as NASA administrator. Though Griffin made it clear he would be willing to remain as administrator under Obama, and though many politicians from both parties wanted him to stay, there were too many issues on which Griffin and Obama disagreed. Griffin was especially firm in his belief that the Ares rocket program should go forward, while the Obama administration seems equally willing to reconsider and possibly cancel it.

The new NASA administrator, Charles Bolden, is a former astronaut who not only flew on the mission that brought Hubble to space in 1990, he was also a member of the 2004 National Academy panel that recommended that NASA fly the shuttle servicing mission to Hubble instead of a robot mission. Bolden is a strong advocate of manned space flight, and his appointment indicates that the interest in human spaceflight by Americans of all political stripes remains strong and unwavering. Whether he will have Mike Griffin's political savvy for focusing and managing the human space program amid the political minefields that routinely litter Washington, D.C., remains unknown. Unlike Griffin, Bolden does not have a very deep or varied management résumé.

The third major event began after the book's publication and is still on-going. As astronomers have done for every decade since the Greenstein Decadal Survey in the 1970s, they are right now in the process of writing a new decadal survey, due out in 2010, to outline their priorities and desires for the next decade.

As part of this process, astronomers have been holding regular meetings, and have also written almost 350 white papers describing what they think the field's goals should be in the next ten years. A review of those white papers reveals some interesting trends, number one of which is an increased desire by astronomers to build a new and much larger optical/ultraviolet space telescope to replace Hubble.

Unlike the 1950s and 1960s, when most astronomers opposed Lyman Spitzer's efforts, the interest of astronomers in the late 2000s in a new space telescope seems strong and growing. Not only have many proposed the need for such a telescope, a number have worked up very detailed proposals, some of which are truly astounding in their audaciousness.

Probably the most ambitious of these proposals is the Advanced Technology Large Aperture Space Telescope (ATLAST). Developed under the leadership of scientists at the Space Telescope Science Institute, the ATLAST proposal outlines in great detail several different telescope designs, including either an 8-meter telescope or a segmented 16.8-meter telescope, both launched on the Ares 5 rocket, or a segmented 9.2-meter telescope launched on an Atlas V or Delta IV rocket. All three options include human or robot servicing capabilities, so that the telescope can be upgraded or repaired and its lifespan extended for several decades after launch.

The largest ATLAST concept, the 16-meter segmented mirror, is especially daring. Almost seven times bigger than Hubble, the telescope and mirror would be launched packed tightly inside the Ares 5 rocket, and then unfold and transform itself once in orbit, much like a robot transformer in a Japanese children's cartoon.

Whether the ATLAST proposals, or any of the other optical space telescopes being put forward at this time, will ever be built is, of course, unknown. That astronomers are now advocating them, however, is strong evidence that they believe a Hubble replacement is necessary. The 2004 cancellation of the last servicing mission to Hubble made astronomers very aware of how close they are to losing their optical and ultraviolet capabilities in space. Without Hubble they will have nothing, and Hubble's future, as bright as it is at this moment, is still limited.

Moreover, the worldwide acclaim and intense interest in the shuttle servicing mission is further evidence that the public cares about this capability as well. Not surprisingly, politicians, recognizing the strength of Hubble's appeal and popularity, also made it a point to show their interest and support, eagerly elbowing their way into Hubble's limelight during the mission. After the repairs and upgrades to Hubble had been completed and the telescope released from the shuttle cargo bay, the astronauts talked with President Obama by phone, then testified from space before a Senate committee hearing, the first time such a thing had ever been done.

As I said earlier in the book, the desire to see the universe clearly is an imperative that cannot be denied. Regardless of what the Obama administration decides to do about NASA's Moon program, it seems that a new optical space telescope is almost certainly in our future.

■ ■ ■

At the end of the last spacewalk during this last servicing mission to Hubble, astronaut John Grunsfeld took a few moments to reflect on Hubble's importance. This was Grunsfeld's third spaceflight and eighth spacewalk to Hubble, and no one had been more passionate or dedicated in his effort to get all of Hubble's repairs and upgrades completed.

"As Arthur C. Clarke says," Grunsfeld reflected, "the only way of finding the limits on the possible is by going beyond them into the impossible."[3]

For most of human history, the furthest range of a person's experience was a distant and unreachable horizon. This untouchable horizon defined "the limits of the possible." No matter how far an individual traveled, there was always a forever-receding horizon line of unknown territory tantalizingly out of reach before him.

In earliest prehistoric times, the size of the known territory within that horizon line was quite small. Each villager knew a region ranging from ten to fifty miles in radius. He or she knew there were people and villages beyond the horizon, but never saw them. Moreover, even the most traveled explorer had a limit to his experience, and knew that at some distant point what lay beyond that horizon was a complete mystery.

Later, as human civilization progressed, the size of known territory within that horizon line expanded. Different cultures met, exchanged information about each other, and recorded this knowledge so that even those who did not travel far from their homes could know something of distant lands beyond their personal horizon. Still, explorers who pushed the horizon found that it continued to forever recede. No matter how far they traveled, the horizon was always ahead of them, an impossible goal beckoning them onward to find new lands unexplored.

Then humans reached the ocean and the sailors took over. To the mariner, there was still a forever-receding horizon of great mystery, which he initially feared traveling out toward because it was dangerous and risky. The ocean was a vast desert, with no food or water. Worse, that desert could suddenly become violent, heaving his ship about and tearing it to shreds.

With time, ship designs improved, and the mariners began journeying outward. The Vikings sailed out into the northern seas to find more lands and more endless unknown territories. Later, using better ships that were more reliable, Columbus pushed the western horizon, and this time the visit was permanent. Even for Columbus, however, the horizon was still an impossible goal, out of reach. He had sailed west, hoping to reach China and thus circumscribe the very limits of the horizon. Instead he discovered the New World, with its vast new territories and unlimited possibilities.

Nonetheless, the impossibility of touching that horizon had never been a deterrence, for either Columbus or anyone else. As the poet Robert Browning wrote, "Ah, but a man's reach should exceed his grasp, / Or what's a heaven for?" People from all cultures felt compelled to reach for that unreachable distant goal, "to go beyond the possible and try to touch the impossible."

It was with Magellan, however, that the impossible became possible and the limits of the horizon were finally reached. For though he set out "to sail beyond the sunset," traveling west as far as he could go, Magellan found that the horizon did *not* recede forever away from him into unknown lands. Instead, the survivors of Magellan's epic voyage circled the globe and found themselves back where they had started, in a known place. The unknown horizon was gone. Humanity for the first time knew the limits of the world. The impossible no longer existed.

For the next five centuries the human race was consumed with learning and exploring and settling the limits of this Earth, delving into its every corner, from the freezing poles to the bottom of the ocean to the highest mountaintop. In all cases, however, the Earth was a prison, placing a curb to exploration. No matter how deeply humans probed, they were still trapped on a spherical world, a giant prison floating in the blackness of space. There was no visible but untouchable horizon to reach for.

The Hubble Space Telescope, along with all the early manned and unmanned space probes of the past half century, has given us back that horizon. We are no longer trapped on this Earth. We can now travel outward with increasing sophistication, either in manned spaceships or with unmanned robots like Hubble, pushing against a new infinite horizon that—instead of a horizon line—is a black sky above us and receding away from us in all directions.

Grunsfeld finished his last spacewalk to Hubble by adding this one small thought: "As Drew [Feustel] and I go into the airlock, I want to wish Hubble its own set of adventures, and with the new set of instruments we've installed, that it may unlock further mysteries of the universe."[4]

Hubble has given us the first detailed and clear glimpse of what lies hidden in the black untouchable horizon above us. May we not shirk from that adventure, but reach out to grasp it fully, even if we cannot ever really touch its limits.

NOTES

CHAPTER 1: FOGGY VISION

1. Burnham, 467; Warner and Warner, 208; Gratton, 308.
2. Bok, 70–72.
3. Gaviola, 408.
4. Ley and von Braun, 51.
5. Ley and von Braun, 62.
6. See, for example, Burnham, 719–722.
7. Spitzer interview, 5/10/1978, 71.
8. Spitzer interview, 6/17/1982, 3–4; Baxter, 33, 408; Smith (1966), 33–39, 46–47n18; Collins, 20–22, 50–54, 120; Spitzer and Ostriker, 369–371.
9. Russell, 112–113; Richardson, 113–123.
10. Spitzer and Ostriker, 372–380.
11. Spitzer interview, 4/8/1977, 1.
12. Spitzer interview, 4/8/1977, 13–15.
13. Spitzer interview, 5/10/1978, 81.
14. Morton (2007), 1; Spitzer interview, 5/10/1978, 77.
15. Doreen Spitzer/Lutie Spitzer Saul interview, 10/5/2006; Spitzer letter to his family, 7/25/1965, Miscellaneous Personal Papers file, Spitzer Papers.
16. *New York Times*, 12/11/1993, 23.
17. Spitzer (1958), 32.
18. Spitzer (1950, 1952, 1956).
19. Spitzer interview, 4/8/1977, 46.
20. Smith (1993), 31.
21. Spitzer interview, 4/8/1977, 51; 3/15/1978, 14; van Allen interview, 7/15/1981, 191–196; Spitzer chronology, Balloon Telescope Miscellaneous file, Spitzer Papers.
22. Schwarzschild interview, 6/3/1977, 38.
23. Trimble, 1291.
24. Schwarzschild interview, 12/16/1977, 102.
25. Spitzer and Ostriker, 515; Spitzer interview, 4/8/1977, 39; 5/10/1978, 60.

26. Schwarzschild interview, 12/16/1977, 102.

27. Schwarzschild interview, 12/16/1977, 102–103; Spitzer interview, 3/15/1978, 1.

28. Schwarzschild interview, 7/19/1979, 160.

29. Schwarzschild interview, 7/19/1979, 161.

30. Spitzer chronology, Balloon Telescope Miscellaneous file, Spitzer Papers; Schwarzschild letter to Spitzer, 9/2/1957, Letters to Department Members 1947–62, Spitzer Papers; Schwarzschild interview, 7/19/1979, 161–162; Rogerson, 112–115; *Sky and Telescope*, "Project Stratoscope Obtains Finest Pictures of the Sun," 79–80; *Science*, "News of Science" (7/10/1959), 89.

31. Newell, 51–57; Dickson, 76–82.

CHAPTER 2: SLOW START

1. Thackeray (1956), 103–105; Thackeray (1961), 160.
2. Geoffrey Burbidge interview, 2/27/2006.
3. Thackeray (1953), 211–236.
4. Geoffrey Burbidge (1962), 304–306; Margaret Burbidge, 547–650.
5. Margaret Burbidge interview, 11/29/1984, 1.
6. Margaret Burbidge interview, 11/29/1984, 2.
7. Roman interview, 8/19/1980, 2–3.
8. Roman interview, 8/19/1980, 15.
9. Letter, Nancy Roman to her parents, 7/12/1953, Nancy Roman Papers.
10. Roman interview, 8/19/1980, 16.
11. Roman interview, 8/19/1980, 18–19.
12. Naugle, 2–3.
13. Newell, 104–105, 499; Naugle, 44–45.
14. Roman interview, 8/19/1980, 22–23.
15. Logsdon, 504–509; Zimmerman (Fall/2002), 14–23.
16. Code interview, 9/30/1982, 28.
17. Code interview, 9/30/1982, 33.
18. Code interview, 9/30/1982, 35–36; Logsdon, 521–525.
19. Roman internal NASA memo, 5/28/1964, AURA Meeting Minutes 1962–64 file, Hubble Space Telescope Collection; Meinel, A.B. "Design Considerations for a Large Aperture Orbital Telescope" report to AURA, February 1969, AURA 1959–61 file, Hubble Space Telescope Collection; Meinel, 369–380.
20. Roman interview, 8/19/1980, 27–28.

21. Tucker and Tucker, 54–55.

22. See Goldberg 1960–75 file in Hubble Space Telescope Collection. See especially Newell, Memorandum for the Files, 3/11/1960; Newell letter to Goldberg, 4/18/1960; Goldberg letter to Newell, 4/22/1960.

23. Newell, 207.

24. Roman interview, 8/19/1980, 50; Goldberg interview, 5/17/1978, 127; Edmondson notebook, June 28–29, in Edmondson—1959–61, 1983–84 file, Hubble Space Telescope Collection.

25. Space Science Board (1962), 2–13; Spitzer letter to Goldberg, 7/20/1962, Space Astronomy: Iowa Study Group 1962 file, Spitzer Papers.

26. Zimmerman (2000), 25–42.

27. Goldberg interview, 5/17/1978, 128.

28. Space Science Board (1965), part 2, 15; see also Spitzer's notes from the meeting, Woods Hole Space Science Study 1965 file, Spitzer Papers.

29. Frederick interview, 5/2/1986, 7.

30. Goldberg interview, 5/17/1978, 137.

31. Roman interview, 2/3/1984, 9.

32. Goldberg interview, 5/16/1978, 5.

33. Goldberg interview, 5/16/1978, 6–8.

34. Goldberg, 5/17/1978, 132.

35. Margaret Burbidge interview, 11/29/1984, 2.

36. Henry interview, 11/29/1983, 20.

37. See Office of Space Science Correspondence 1958–61 file and OSSA Correspondence 1959–97 file, NASA History Office; see especially memos, 6/9/1966, 12/21/1966, 1/17/1967, 1/31/1967, 2/27/1967, 3/7/1967, 3/15/1967, 3/27/1967, 4/3/1967, 4/5/1967, 4/20/1967, 4/25/1967, 6/12/1967, 6/23/1967; also, Henry Smith letter to Spitzer, 2/6/1967; Schwarzschild letter to Smith, 3/24/1967; Schwarzschild letter to Newell, 7/6/1967.

38. Minutes, Astronomy Missions Board, Hubble Space Telescope Collection.

39. O'Dell interview, 1/10/2006; 4/21/2006.

40. O'Dell interview, 1/10/2006.

41. O'Dell interview, 4/12/1982, 3.

42. O'Dell interview, 4/12/1982, 1–14; 4/21/2006.

43. O'Dell, 4/12/1982, 17.

44. O'Dell interview, 5/21/1985, 34.

45. O'Dell interview, 4/12/1982, 13.

46. Minutes, Astronomy Missions Board, Hubble Space Telescope Collection. See in particular minutes of 10th meeting, October 10–11, 1968.

47. O'Dell interview, 4/12/1982, 17.

48. O'Dell interview, 4/12/1982, 17; Stuhlinger and Ordway, vol. 1, 250, vol. 2, xiv; Freeman, 124; Ward, 31–42; Piszkiewicz, 31–38.

49. O'Dell interview, 4/12/1982, 18–19.

50. *Optical Telescope Technology*; Smith (1993), 68–69, 72–80; Marshall Space Flight Center, *Large Space Telescope Phase A Final Report*, vols. 1–5.

51. Bless interview, 11/3/1983, 23–24; Brandt interview, 5/15/1989, 11; Leckrone interview, 8/14/1984, 12–13; Mitchell interview, 12/10/1984, 28–30; Emanuel interview, 12/2/1985, 17–18; Smith (1993), 82–83.

52. Goldberg interview, 2/22/1983, 45–49.

53. Schwarzschild interview, 6/3/1977, 65; email, Lutie Spitzer Saul, 2/15/2007.

54. Schwarzschild interview, 6/18/1982, 15.

55. Stuhlinger interview, 4/2/1984, 3; Spitzer interview, 10/27/1983, 11–12; O'Dell interview, 4/12/1982, 21–23; 4/21/2006.

56. Mitchell interview, 12/10/1984, 28–30.

CHAPTER 3: GETTING MONEY

1. Morton (2007), 2–3; Morton interview, 1/5/2006; Spitzer letter to his family, 7/25/1965, Miscellaneous Personal Papers file, Spitzer Papers.

2. Morton (1993), 4; Morton (2007), 4–5.

3. Bohlin interview, 1/4/2006.

4. Bless interview, 11/3/1983, 24.

5. O'Dell interview, 4/21/2006; memo from James T. Murphy to Dr. Rees, 11/2/1972, O'Dell memo to James T. Murphy, 11/6/1972, both in Reading file, 1972–73, Hubble Space Telescope Collection; letter from George Pieper to Kenneth Hallam, 12/6/1972, Chronological Correspondence 1972–76 file, Hubble Space Telescope Collection.

6. Memo, 12/22/1972, Misc. Papers 1971–82 file, Hubble Space Telescope Collection.

7. O'Dell interview, 5/21/1985, 7–8.

8. *Hearings before the Committee on Aeronautical and Space Sciences*, Senate, 2/2/1974, 20.

9. *Hearings before the Subcommittee on Appropriations*, House, 3/26/1974, 26.

10. *Hearings before the Subcommittee on Appropriations*, House, 3/26/1974, 248–249.

11. *Hearings before the Subcommittee on Appropriations*, House, 3/26/1974, 327–330.

12. Greenstein, vol. 1, 8–9.

13. *Hearings before the Subcommittee on Appropriations*, House, 3/27/1974, 742–745.

14. *House Reports*, vols. 1–5, Report 93–1139, June 21, 1974.

15. *Congressional Record*, 7/26/1974, 21220–21247.

16. O'Dell Daily Notes, 6/20/1974, Hubble Space Telescope Collection.

17. O'Dell interview, 4/21/2006.

18. O'Dell interview, 4/21/2006; O'Dell Daily Notes, 6/20–28/1974, Hubble Space Telescope Collection.

19. Greenstein interview, 7/21/1977, 73–74.

20. Spitzer and Ostriker, 379; Spitzer interview, 3/7/1984, 11.

21. Greenstein interview, 5/19/1978, 235.

22. Greenstein, 8; Kellerman and Moran, 470–475.

23. Greenstein interview, 5/19/1978, 235.

24. O'Dell Daily Notes, 6/21/1974, Hubble Space Telescope Collection; Greenstein interview, 5/19/1978, 231–234.

25. Regis, 5–6.

26. Neta Bahcall interview, 10/5/2006; Regis, 155–164.

27. O'Dell interview, 5/21/1985, 15.

28. O'Dell interview, 5/21/1985, 15; O'Dell Daily Notes, 4/25/1973, Hubble Space Telescope Collection.

29. John Bahcall interview, 11/3/1983, 5.

30. O'Dell interview, 5/21/1985, 15.

31. Hargittai and Hargittai, 233–235; Nathan interview, 11/6/2006.

32. Regis, 219–226; Zimmerman (1996), 16–19.

33. Regis, 171; Neta Bahcall interview, 10/5/2006.

34. See letters by Spitzer to various elected officials, Large Space Telescope and Congress: Activities L. Spitzer 1974 file; also General Correspondence, Representatives and Senators, Spitzer Papers.

35. Telegram to Hinners, 8/8/1974, Large Space Telescope and Congress: Activities L. Spitzer 1974 file, Spitzer Papers; Smith (1993), 133.

36. *Congressional Record*, 7/2/1974, 22088–22089; Smith (1993), 134.

37. O'Dell Daily Notes, 7/2/1974, 7/5/1974, 7/7/1974, 7/9/1974, 7/11/1974, 7/16/1974, 7/26/1974, Hubble Space Telescope Collection.

38. *Washington Post*, 7/4/1974, A14; 8/2/1974, A11; 8/9/1974, A10; 8/15/1974, C1–C2; 9/11/1974, A1; 9/12/1974, B1; 11/2/1974, B1; 11/3/1974, A1; 11/8/1974, A2, A26.

39. John Bahcall interview, 12/20/1983, 8.

40. John Bahcall interview, 12/20/1983, 11–12.

41. Spitzer interview, 6/17/1982, 62.

42. *Congressional Record*, 8/5/1974, 26745.

43. *Senate Reports, 13057–6, 93rd Congress, 2nd Session*, Senate Report 93–1091, August 15, 1974, 19.

44. O'Dell Daily Notes, 8/2/1974, Hubble Space Telescope Collection; Spitzer letter, 8/7/1974, LST 1974–75 file, NASA History Office.

45. O'Dell interview, 4/21/2006; O'Dell Daily Notes, 8/29/1974, 9/2/1974, Hubble Space Telescope Collection.

46. Minutes, LST Operations and Management Working Group, 12/13/1974, 5, Hubble Space Telescope Collection.

47. Smith (1993), 160–163.

48. Hinners interview, 10/17/1984, 16–18.

49. O'Dell interview, 4/21/2006; John Bahcall letter to Spitzer, 12/22/1975, Large Space Telescope and Congress: Budget Considerations 1976 file, Spitzer Papers; Chronology, 1976–77, Large Space Telescope and Congress: Budget Approval, 1977, Spitzer Papers.

50. Joint letter to Fletcher, 1/21/1976, Spitzer file, Hubble Space Telescope Collection; John Bahcall letter to Fletcher, 1/22/1976, Congressional Committees 1976 file, Hubble Space Telescope Collection.

51. John Bahcall interview, 11/3/1983, 11–13; Bahcall letter to Spitzer, 1/9/1976, Large Space Telescope and Congress—Budget Considerations 1976 file, Spitzer Papers.

52. O'Dell letter to E. N. Parker, 12/23/1974, O'Dell Reading file 1974–75, Hubble Space Telescope Collection.

53. O'Dell interview, 4/21/2006; O'Dell letter to George Pieper, 1/7/1976, O'Dell Reading file 1976–77, Hubble Space Telescope Collection; O'Dell Daily Notes, 12/24/1975, O'Dell Papers.

54. John Bahcall interview, 3/22/1984, 1; O'Dell Daily Notes, 4/1/1976, 4/14/1976, 4/15/1976, O'Dell Papers.

55. John Bahcall interview, 11/3/1983, 22; Wallerstein letter, circa 2/1976, Congressional Committee 1976 file, Hubble Space Telescope Collection; O'Dell Daily Notes, 1/15/1976, O'Dell Papers.

56. Smith (1993), 170; Field interview, 7/14/1980, 6; 3/10/1986, 21–22.

57. *Hearings before a Subcommittee of the Committee on Appropriations*, February 18, 1976, 134–148; Smith (1993), 173.

58. Smith (1993), 181; Field interview, 3/10/1986, 13–14, 25; Field letter to Bahcall, 2/25/1976, Large Space Telescope and Congress-Budget Considerations 1976, Spitzer Papers.

59. Minutes, Management and Operations Working Group in Shuttle Astronomy, February 17–18, 1976, 9–10, MOWG Shuttle Astronomy 1974–76 file, Hubble Space Telescope Collection; O'Dell Daily Notes, 2/6/1976, O'Dell Papers.

60. Field interview, 3/10/1986, 16; John Bahcall interview, 12/20/1983, 19–20; O'Dell Daily Notes, 2/3/1976, 3/2/1976, 3/4/1976, O'Dell Papers.

61. Bahcall letter to George Field, 6/4/1976, Congressional Legislative 1974–76 file, Hubble Space Telescope Collection; O'Dell Daily Notes, 4/15/1976, 4/16/1976, 4/27/1976, 6/3/1976, O'Dell Papers.

62. *House Reports 13134–8*, 94th Congress, 2nd session, Report 94–1220, 26.

63. Myron Smith letter to O'Dell, 7/12/1976, O'Dell Reading file 1976–77, Hubble Space Telescope Collection.

64. *Congressional Record*, 7/22/1976, 23538–23539; 7/27/1976, 24046–24053.

65. John Bahcall memo, 7/29/1976, Congressional Legislative 1974–76 file, Hubble Space Telescope Collection; Smith (1993), 180.

66. Margaret Burbidge interview, 11/29/1984, 5–7.

67. John Bahcall interview, 12/20/1983, 13; transcript extracts of John Bahcall's interview Dailies, circa Spring 1985, for TV series *Smithsonian World*, 42, Hubble Space Telescope Collection; *Hearings Before a Subcommittee of the Committee on Appropriations*, March 29, 1977, 100–101.

68. Proxmire press release, 5/11/1977, in ST New Start 1976–78 file, Hubble Space Telescope Collection.

69. Smith (1993), 185; *Congressional Record*, 7/19/1977, 23668–78; *Senate Reports, 13168–5, 95th Congress, 1st session*, Senate Report 95–280, June 21, 1977, 42.

70. Smith (1993), 225.

CHAPTER 4: BUILDING IT

1. Westphal interview, 8/12/1982, 160–171; Westphal et al., L95–L98.

2. Kris Davidson interview, 9/19/2006; van Genderen et al., 317–373; Warren-Smith et al., 761–768.

3. Roman interview, 2/3/1984, 27.

4. Leckrone interview, 8/14/1984, 31.

5. Sobieski and Wampler, 59–62, 83–84; Spitzer and Zabriskie, 412–420; Smith (1993), 107–109; Smith and Tatarewicz, 1125–1127.

6. Hinners interview, 11/16/1984, 12.

7. John Bahcall interview, 12/20/1983, 30–32; 3/28/1984, 9–12; Sobieski interview, 2/8/1985, 15–16; Lowrance interview, 3/28/1985, 18–25; Minutes,

ST Operations and Management Working Group, 10/18–19/1976, 4–7, Hubble Space Telescope Collection.

8. Westphal interview, 8/9/1982, 13.

9. Westphal interview, 8/9/1982, 11; *New York Times*, 9/14/2004, C23.

10. Westphal interview, 8/9/1982, 30.

11. Westphal interview, 8/9/1982, 48–52; Personal communication, David DeVorkin, 8/29/2006.

12. Westphal interview, 8/12/1982, 148.

13. Westphal interview, 8/12/1982, 182–184.

14. Westphal interview, 8/12/1982, 186–188; O'Dell interview, 4/21/2006.

15. Westphal interview, 8/12/1982, 193–198, 208–209.

16. O'Dell interview, 5/21/1985, 25; O'Dell letter to John Bahcall, 10/3/1977, O'Dell reading file 1976–77, Hubble Space Telescope Collection.

17. Spitzer interview, 5/10/1978, 95; 6/17/1982, 73; Lowrance interview, 3/28/1985, 18.

18. Downey memo, 6/3/1966, Hubble Space Telescope Collection. Downey notes that this was not a direct quote, but "the gist" of what Friedman said.

19. O'Dell interview, 4/21/2006; O'Dell lecture, Johns Hopkins University, 4/21/2006.

20. Minutes, 7th meeting, LST-WG, 10/1–2/1974, 6–7, Hubble Space Telescope Collection

21. Minutes, ST Operations and Management Working Group, 10/18–19/1976, 2–3, Hubble Space Telescope Collection; Smith (1993), 207–211.

22. Minutes, 7th meeting, LST-WG, 10/1–2/1974, 6–7, Hubble Space Telescope Collection; Minutes, LST Operations and Management Working Group, 4/3–4/1975, 5, Hubble Space Telescope Collection.

23. Davidsen interview, 1/29/1984, 10–12; Teem interview, 9/26/1984, 11; *Science*, 8/5/1983, 534–35.

24. Teem interview, 9/26/1984, 16; John Bahcall interview, 3/22/1984, 19.

25. John Bahcall interview, 3/22/1984, 24–26; Teem interview, 9/26/1984, 16.

26. Spitzer interview, 3/7/1984, 32.

27. O'Dell interview, 12/15/1982, 13.

28. John Bahcall interview, 3/22/1984, 15–28; Davidsen interview, 1/29/1984, 10–20.

29. Spitzer interview, 3/7/1984, 32–33.

30. Giacconi interview, 2/23/2006; O'Dell Daily Notes, 5/9/1979, Hubble Space Telescope Collection; Stofan letter to Teem, 6/12/1981, STScI 1980–81 file, Hubble Space Telescope Collection.

31. Keathley memo to Lucas, 4/16/1979, Notes to Lucas 1978–80 file, Hubble Space Telescope Collection; O'Dell Daily Notes, 11/13/1979, Hubble Space Telescope Collection; Wood Daily Notes, 4/14/1992.

32. Capers and Lipton, 4.

33. Brandt letter, 8/3/1979, Space Telescope 1979 file; O'Dell Daily Notes, 4/23–25/1979; Keathley letter to Lucas, 4/19/1979, Notes to Dr. Lucas, 1978–80 file; all in Hubble Space Telescope Collection.

34. O'Dell interview, 5/21/1985, 54.

35. Speer interview, 2/26/1985, 1–5.

36. available as of 3/28/2006 at http://www.nasm.edu/research/aero/trophy/laureate/1977.htm and http://www.nasm.edu/research/aero/trophy/avweek.htm.

37. O'Dell interview, 5/21/1985, 60; O'Dell Daily Notes, 1/24/1980, 1/31/1980; O'Dell letter to Lucas, 1/23/1980, reading file 1980, Hubble Space Telescope Collection.

38. O'Dell Daily Notes, 3/21/1980, Hubble Space Telescope Collection; Emanuel interview, 30–32.

39. O'Dell interview, 5/21/1985, 54.

40. O'Dell interview, 5/21/1985, 59–60; Smith (1993), 274–275; Fastie letter to O'Dell, 9/17/1981, Marshall Robert Smith Notes 1981 file, Hubble Space Telescope Collection.

41. O'Dell interview, 5/21/1985, 34, 66.

42. O'Dell interview, 4/21/2006; O'Dell Daily notes, 3/21/1980, 4/4/1980, 4/23/1980, Hubble Space Telescope Collection.

43. Smith (1993), 256; Brandt interview, 5/15/1989, 7.

44. Weiler interview, 3/17/1986, 23; Brandt interview, 5/15/1989, 17; Brandt letter, 8/3/1979, Space Telescope 1979 file, Hubble Space Telescope Collection.

45. O'Dell interview, 4/21/2006; O'Dell Daily Notes, 4/23/1980, Hubble Space Telescope Collection.

46. Minutes, Science Working Group, May 5–6, 1980, 44, Hubble Space Telescope Collection.

47. O'Dell Daily Notes, 5/7/1980, Hubble Space Telescope Collection.

48. O'Dell memo to Speer, 6/17/1980, ST 1980 file, Hubble Space Telescope Collection.

49. Leckrone memo to Speer, 6/30/1980, ST 1980 file, Hubble Space Telescope Collection.

50. O'Dell interview, 5/21/1985, 58.

51. Kelsall memo, 7/18/1978, Robert Smith Notes 1978, Hubble Space Telescope Collection; Levin trip report, 1/15/1979, Space Telescope 1979, Hubble Space Telescope Collection; O'Dell Daily Notes, 1/11/1979, 1/12/1979, Hubble Space Telescope Collection; Margaret Burbidge interview, 11/29/1984, 12–16.

52. O'Dell letter to John Bahcall, 10/3/1977; Doxsey interview, 7/22/1987, 11.

53. Pellerin interview, 8/1/1983, 19; Leckrone interview, 8/14/1984, 15, 51.

54. O'Dell Daily Notes, 6/19/1980, Hubble Space Telescope Collection.

55. O'Dell Daily Notes, 9/7/1980, Hubble Space Telescope Collection; Minutes, Science Working Group, 9/8–9/1980, 5–8, Hubble Space Telescope Collection.

56. O'Dell interview, 5/21/1985, 34, 66; O'Dell Daily Notes, 10/3/1980, Hubble Space Telescope Collection.

57. O'Dell Daily Notes, 8/7/1980, Hubble Space Telescope Collection.

58. John Bahcall letter to O'Dell, 7/30/1980, Cost Cutting Activity 1980 file, Hubble Space Telescope Collection.

59. O'Dell letter to Bless, 9/16/1980, O'Dell reading file 1980, Hubble Space Telescope Collection.

60. O'Dell Daily Notes, 5/12/1981, Hubble Space Telescope Collection.

61. O'Dell interview, 5/21/1985, 70; O'Dell Daily Notes, 5/21/1981, 8/18–20/1981, 11/24/1981, 12/11/1981, 2/18/1982, 3/2/1982, 3/9/1982, 5/18/1982, 5/25–26/1982, 9/1/1982, Hubble Space Telescope Collection.

62. O'Dell letter to Marc Aucremanne, NASA headquarters, 2/10/1975, Rosendahl 1975–76 file, Hubble Space Telescope Collection.

63. Capers and Lipton, 52.

64. O'Dell Daily Notes, 8/5/1980, Hubble Space Telescope Collection; Capers and Lipton, 1–24; Dunkle, 1985, also available in revised and more complete form at http://www.terrydunkle.com/glass.htm as of 8/8/2006.

65. O'Dell Daily Notes, 7/17/1981, Hubble Space Telescope Collection; Capers and Lipton, 44.

66. Minutes, Science Working Group meeting, 10/26–27/1982, 2, Hubble Space Telescope Collection; Speer letter to Rehnberg, 10/26/1982, Project Manager reading file 10/1982, Hubble Space Telescope Collections.

67. Presentation to Beggs, 11/3/1982, Hubble Space Telescope Collection; Weiler interview, 10/20/1983, 14.

68. Fordyce interview, 10/31/1983, 9–10, 18–19, 25; Capers and Lipton, 1–24.

69. Smith (1993), 301–305; Pellerin interview, 8/1/1983, 5–10.

70. Lucas letter to Beggs, 4/8/1983, Program Manager reading file 1983, Hubble Space Telescope Collection.

71. Bless interview, 11/3/1983, 31; Brandt interview, 5/15/1989, 23–24, 28–30; Richards interview, 2/20/1984, 42–44; Guha interview, 8/1/1983, 44–45; Reetz interview, 9/21/1983, 25–26.

72. O'Dell interview, 4/21/2006.

73. Capers and Lipton, 1–24; Allen, sections 5–1, 6–1, 7–1, 8–1, 9–1; O'Dell lecture, 4/21/2006, Johns Hopkins University.

74. Vaughan, 175.

75. Olivier interview, 1/8/1987, 33; Welch interview, 2/12/1987, 17–19.

CHAPTER 5: SAVING IT

1. Lauer interview, 5/18/2006; Holtzman interview, 5/23/2006; Burrows interview, 5/25/2006; Smith (1993), 402–403.

2. Westphal interview, 9/14/1982, 254–257.

3. Foster et al., 634–644.

4. Stockman interview, 2/15/2006; Burrows interview, 5/25/2006.

5. *New York Times*, 6/4/1948, 1; 6/5/1948, 1; Christianson, 314–315.

6. Brad Smith letter to Jim Westphal, 8/7/1979, Chronological Correspondence 1979 file, Hubble Space Telescope Collection.

7. Weiler interview, 10/17/2006; Leckrone interview, 10/23/2006; Leckrone, personal communication, 10/23/2006; NASA Level I Policies (revised), provided by Leckrone, 10/23/2006.

8. Westphal interview, 9/28/1985, 21–29; Neta Bahcall interview, 10/5/2006; Weiler interview, 10/17/2006.

9. Minutes, Science Working Group Meeting, 1/14–15/1986; Minutes, Science Working Group Meeting, 4/2–3/1986; Robert Smith's notes of Science Working Group Meeting, 4/3/1986, 7–8, Hubble Space Telescope Collection; Westphal interview, 9/14/1982, 258–259.

10. Burnham, 1383; Space Telescope Science Institute, Hubble Space Telescope: GO and GTO Observing Program, 1990, 9–12.

11. Villard, personal communication, 4/17/2006; Villard interview, 8/23/2006; Weiler interview, 10/17/2006; Chaisson, 97–102. It should be noted that Chaisson is somewhat unreliable. Weiler and Villard (who described Chaisson's

book to me as "a docudrama") were both eyewitnesses to these events however and did confirm them.

12. Prelaunch press conference, 4/9/1990, Space Telescope Science Institute Archives.

13. First light press briefing, 5/20/1990, Space Telescope Science Institute Archives.

14. Holtzman interview, 5/23/2006; Lauer interview, 5/18/2006, 9:00–14:00 minutes; Smith (1993), 403.

15. Space Telescope Science Institute, *Hubble Space Telescope: Optical Telescope Assembly Handbook*, 1990, 1.

16. Burrows interview, 5/25/2006.

17. Schroeder, 43–52; Harrington, 12, 18–19; Thompson, 182–184.

18. Faber interview, 7/31/2002, 101–113; Dressler, 1995, 238–264.

19. Westphal interview, 9/28/1985, 8, 18–21; Faber interview, 4/18/2006; also Faber, personal communication, 4/18/2006.

20. Smith (1993), 406.

21. Faber memo to Westphal, 6/8/1990, Faber Notebook.

22. Westphal interview, 9/28/1985, 8–10.

23. Faber interview, 4/18/2006; Faber notebook, 6/27/1990; Smith (1993), 415; Chaisson, 181.

24. John Bahcall interview, 8/26/1992, 10.

25. O'Dell, personal communication, 8/21/2006.

26. Weiler interview, 9/13/2006; 2/11/1986, 14–15; 3/17/1986, 1–3; Minutes, Science Working Group, 6/17–19/1981, 1; 6/29–30/1982, 3; Minutes, STOPAT, Meeting #3, 8/26/1983, 2; STOPAT meeting #8, 1/16/1984, 3; Minutes, Science Working Group, 2/1–2/1984, 2; In the same file: Notes taken by Joe Tatarewicz at Science Working Group meeting, 2/1/1984, 7–9 and Weiler White Paper, enclosure #3; Minutes, Science Working Group, 6/5–6/1984, 1; Robert Smith's notes of 6/5–6/1984 meeting, 4; Minutes, Science Working Group, 1/14–15/1986, 4–5; all in Hubble Space Telescope Collection.

27. Thompson interview, 6/5/1992, 3–10.

28. Press Conference, 6/27/1990, transcript, 10, Space Telescope Science Institute Archives.

29. *Washington Post*, 6/28/1990, A1; *St. Louis Post-Dispatch*, 6/28/1990, 16A; *Baltimore Sun*, 6/28/1990, 1A; *Chicago Tribune*, 6/28/1990, 1; *New York Times*, 6/28/1990, A1; 6/29/1990, A1; *Los Angeles Times*, 6/30/1990, A1, A37, B6; *Miami Herald*, 6/28/1990, 1A;

30. Allen, iii-iv, 7–1 to 7–7; Wood Daily Notes, 5/30/1991, 5/31/1991.

31. Giacconi interview, 2/23/2006; Ford interview, 10/13/2006.

32. Keller interview, 2/22/1985, 17; Smith notes of Science Working Group meeting, 10/8–9/1985, 10, Hubble Space Telescope Collection; Giacconi interview, 2/23/2006; Teem interview, 10/4/1984, 10–18, 31–32.

33. Giacconi interview, 1/25/1984, 24–26, 42–43; Schreier interview, 7/6/1987, 7–24.

34. Giacconi interview, 1/25/1984, 42.

35. Faber notebook, 9/3/1990; Crocker interview, 9/15/2006.

36. Crocker interview, 9/15/2006.

37. Weiler interview, 3/9/1987, 14–19; Brown interview, 6/7/1984, 1–23; Leckrone interview, 8/14/1984, 26–28.

38. Pellerin unpublished article, "The Hubble Experience," 9–10; Pellerin, personal communication, 8/26/2005.

39. Loomis presentation, 12/5/1990, 1990 Costar file, Rodger Doxsey files; Crocker interview, 9/15/2006; Ball Aerospace and Communications Group, 11–13; Crocker, 22.

40. Wood Daily Notes, 9/17/1992; Schuiling (1994), 81; see also http://www.astronautix.com/flights/sts61.htm, available as of 8/21/2006.

41. Wood Daily Notes, 8/14/1991; 8/16/1991; 12/16/1991.

CHAPTER 6: "NEW PHENOMENA NOT YET IMAGINED"

1. Hester interview, 9/18/2006; Hester et al., 654–657, 835.

2. Video of "Second Light," 12/17/1993, Space Telescope Science Institute Archives; Burrows interview, 5/25/2006; Hester interview, 6/14/2005; Weiler and Jensen, 18.

3. Hester interview, 6/14/2005.

4. Zimmerman (3–4/1994), 8.

5. Nelson, 1.2, 4.4–4.7.

6. Smith and Tatarewicz, 1221–1237; *Washington Times*, 7/27/2003, available as of 8/30/2006 at http://washingtontimes.com/business/20030727-104257-3030r.htm.

7. Beckwith testimony, Hubble Panel hearing, 6/22/2004.

8. Giacconi interview, 2/23/2006.

9. Bahcall et al., 642–658; Hutchings and Neff, 1–14.

10. Livio et al., 64–77.

11. Fishman, 467–476; Ruderman, 164–180; Paczynski, 321–330; Odewahn et al., 509:L5–L8; Bloom et al., 507:L25–L28; Zimmerman (1–2/2000).

12. Hubble, 90–101.

13. Christianson, 189–195.

14. Livio et al., 214–221.

15. Kirshner, 194–261; Perlmutter et al., 51–54; Glanz, 44–51; Livio, 42–49; Regis, 180; Gamow, 44.

16. Sembach et al., 320; Zimmerman (1998), 13–16.

17. Hester interview, 6/14/2005; Weiler interview, 9/13/2006; Lauer, personal communication, 10/19/2006.

18. Giacconi interview, 2/23/2006.

CHAPTER 7: ABANDONMENT

1. *Columbia Accident Investigation Board Report*, vol. 1, 9, 100, 117–118, 172, 195–212.

2. Sietzen and Cowing, 114–118; personal communication with Frank Sietzen, 3/16/2006.

3. O'Keefe testimony, Hubble Panel hearing, 6/22/2004.

4. Information about the roadmap process was available as of 10/13/2006 at http://www.hq.nasa.gov/office/apio/apio_strat_roadmaps.htm.

5. Leidecker and Thomas, 5–6.

6. O'Keefe testimony, House Science Committee, 9/10/2003, available as of 10/13/2006 at http://commdocs.house.gov/committees/science/hsy89217.000/hsy89217_0.HTM.

7. Dick, paper and PowerPoint presentation.

8. Bely letter to Riccardo Giacconi and Peter Stockman, 3/26/1985, Stockman Papers.

9. Bely and Burrows, 3–6, 117–160, 177–182.

10. Dressler, 1996, 21–26; *The Next Generation Space Telescope* (1998), 5–39.

11. Alan Dressler letter to Peter Stockman, 1/24/1994, Stockman Papers.

12. Weiler interview, 9/13/2006; Stockman interview, 2/15/2006; Dressler, 1996, 23–24; Stockman email, 8/17/1995, Stockman Papers; Dressler letter, 11/14/1994, Stockman Papers; Spitzer notes, STIC Meeting, 10/19/1994, 3, STIC Agenda Minutes 1994, Spitzer Papers.

13. Beckwith interview, 5/26/2004.

14. *The Next Generation Space Telescope* (1998), 13–14.

15. Dressler (1996), 23.

16. Bahcall (2003), 7.

17. Dick, paper and PowerPoint presentation.

18. Sietzen and Cowing, 160.

19. *Washington Post*, 1/15/2004, A1, A10; Sietzen and Cowing, 171–172.

20. Beckwith memo, 1/16/2004, available as of 10/13/2006 at http://www.stsci.edu/resources/sm4meeting.html; Sietzen and Cowing, 169–173.

21. *Denver Post*, 3/22/2004.

22. Beckwith lecture, Space Telescope Science Institute, 3/16/2006, 20:00–26:00 minutes, available as of 10/13/2006 at http://www.stsci.edu/institute/sd/meridian.html.

23. *Hearing of the Veterans Affairs and House and Urban Development and Independent Agencies Subcommittee of the Appropriations Committee*, 3/11/2004, 22–26; Bond/Mikulski letter to O'Keefe, 3/11/2004; Bond/Mikulski letter to Walker, Comptroller General of the U.S., General Accounting Office, 3/11/2004; Mikulski letter to O'Keefe, 3/11/2004.

24. Faber interview, 4/18/2006.

25. Crocker interview, 9/15/2006.

26. Cepollina interview, 10/2/2006; Cepollina testimony and PowerPoint presentation, Hubble Panel hearings, June 2–3, 2004; Timeline for HRSDM (Hubble Robot Servicing Deorbit Mission), provided by Cepollina, 10/5/2006; DVD of robot simulations, provided by Cepollina, 10/10/2006.

27. Press release, 6/1/2004, available as of 10/13/2006 at http://www.nasa.gov/home/hqnews/2004/jun/HQ_04173_hubble_robotic.html.

28. O'Keefe testimony, Hubble Panel, 6/22/2004; Faber notebook, 6/22/2004, 6.

29. *New York Times*, 7/23/2004, A19; *Washington Post*, 9/8/2004, A21; 9/9/2004, A3; 9/27/2004, A2; 10/2/2004, A2.

30. Faber notebook, 6/22–24/2006, 13.

31. Committee on the Assessment of Options for Extending the Life of the Hubble Space Telescope. *Letter Report, July 13, 2004*, 4, 6.

32. Committee on the Assessment of Options for Extending the Life of the Hubble Space Telescope, *Assessment of Options for Extending the Life of the Hubble Space Telescope, Final Report*, 6, 7, 16.

33. Budget press briefing with NASA comptroller Steve Isakowitz, 2/4/2005; author's personal files.

34. See http://www.nasa.gov/about/highlights/griffin_bio.html, available as of 10/13/2006.

35. Michael Griffin's statement at Goddard, 10/31/2006; press conference at Goddard, 10/31/2006.

CHAPTER 8: THE LURE OF THE UNKNOWN

1. O'Dell interview, 4/21/2006; Leckrone interview, 10/23/2006; O'Dell personal communication, 10/27/2006; Spitzer notes of meeting, 4, Space Telescope Science Institute Working Group 1984 file; Spitzer notes of O'Dell phone conversation, 2/3/1984, L. Spitzer/O'Dell HRS Program Proposal, 1984–1988–1991 file, Spitzer Papers; Space Telescope Science Institute, *The GO and GTO Observing Programs, Version 1.0*, 32–34.

2. Simmons interview, 9/10/1984, 27–29, 34–37, 40–55; O'Dell interview 5/21/1985, 6–7, 82–83; O'Dell Daily Notes, 5/14/1973, 5/24/1973, 6/18/1973, 8/30/1973, 6/25/1974, 6/28/1974, 7/2/1974, 7/10/1974, Hubble Space Telescope Collection; Simmons, 1–3, 15–18, 19–22; Smith (1993), 120–123.

3. Roman interview, 2/3/1984, 15–17; Rosendhal interview, 12/23/1985, 21–22; King interview, 7/20/1978, 44; King interview, 5/2/1984, 3–4; Weiler interview, 9/13/2006; *Outlook for Space: Report*; *Outlook for Space: A Synopsis.*

4. Lowrance interview 3/28/1985, 3–4, 28–30; Schwarzschild interview, 7/19/1979, 158, 163, 167; Bahcall interview, 3/28/1984, 11; Minutes, Science Working Group, 5/5–6/1976, 1; Smith and Tatarewicz, 1227–1231; Longair and Warner, 3.

5. Olivier interview, 1/18/1984, 1–7, also interview memo, 1–13; 1/8/1987, 1–5; Odom interview, 2/26/1985, 31; Downey interview, 1/18/1984, 1–15; Nein interview, 6/5/1984, 4, 21.

6. Minutes, Science Working Group 6/5–6/1984, including O'Dell memo to Burt Edelson, 4/9/1984, Brown memo, 5/11/1984, Smith notes of meeting, 7, Hubble Space Telescope Collection; O'Dell Daily Notes, 4/24/1984; O'Dell interview, 1/10/2006.

7. Spitzer interview, 5/10/1978, 96.

8. Richards interview, 2/20/1984, 48–52; personal communications with Richards, 11/22/2006.

9. See http://www.seds.org/spider/oaos/oaos.html as of 11/16/2006.

10. Committee on the Assessment of Options for Extending the Life of the Hubble Space Telescope, *Assessment of Options for Extending the Life of the Hubble Space Telescope, Final Report*, 27.

11. This was said to me by William Hatfield at a meeting of his local amateur radio club in the spring of 2006.

12. See the talks and presentation material given at the Astrophysics Enabled by the Return to the Moon Conference, held 11/28–30/2006 at the

Space Telescope Science Institute, Baltimore, Maryland, and available as of 12/4/2006 at http://www.stsci.edu/institute/center/information/streaming/archive/AERM.

13. *Space News*, 10/25/2004, 13.

14. Spitzer interview, 6/17/1982, 40–42; 10/27/1983, 45–47; Cox interview, 1/3/2006; Lutie Spitzer Saul, personal communication, 11/5/2006; See also http://www.americanalpineclub.org/pages/page/49, available as of 11/4/2006; Zimmerman (2000), 107–108.

AFTERWORD TO THE PAPERBACK EDITION

1. Harwood (May 14, 2009).
2. Harwood (May 17, 2009).
3. Harwood (May 18, 2009).
4. Harwood (May 18, 2009).

BIBLIOGRAPHY

Of my primary sources, I relied most heavily on the Hubble Space Telescope Collection at the Milton S. Eisenhower Library at Johns Hopkins University. In 1990 Robert Smith wrote *The Space Telescope: A Study of NASA, Science, Technology, and Politics*, a detailed academic description of the telescope's history from inception through launch. The notes and files he compiled for that history he then donated to the Eisenhower Library, where they have been carefully catalogued as the Hubble Space Telescope Collection. In addition, the interviews he and his colleagues conducted for that history are held in the archives of the National Air and Space Museum as part of the Space Telescope History Project. Both archives provided a fundamental foundation for my research.

Other archives of importance included the Lyman Spitzer Papers at the Princeton University Library in Princeton, New Jersey, the Nancy Roman Papers in the Niels Bohr Library at the American Institute of Physics in College Park, Maryland, and the NASA History Office Archives, NASA Headquarters, Washington, DC.

I also used original materials supplied to me by several important individuals in Hubble's history. Sandy Faber provided me an annotated copy of the notebook she kept from May to July, 1990, the daily notes she kept in 2003 as a member of the Committee on the Assessment of Options for Extending the Life of the Hubble Space Telescope, and an annotated copy of Eric Chaisson's book, *The Hubble Wars*. (As I mentioned in one endnote, Chaisson's book contains a great deal of important material, but his reporting of events is very unreliable and must always be checked.) She also provided me many of her files and viewgraph presentations from the 1990 period.

Though a copy of Bob O'Dell's Daily Notes are in the Hubble Space Telescope Collection at Johns Hopkins, the archive lacks the years from

1975 to 1977. Bob O'Dell was kind enough to provide me copies of his Daily Notes from those years, relating to the lobbying effort to fund Hubble.

John Wood at the Goddard Space Flight Center, who was closely involved in the effort to repair Hubble, allowed me to review and copy his daily notes covering the period from 1990 to 1994. Charles Pellerin and Ted Cox provided me their written, unpublished reminiscences, while Rodger Doxsey, Peter Stockman, and Ray Villard of the Space Telescope Science Institute allowed me to review and copy extensive materials from their files. Stephen Beckwith was also kind enough to provide me the visual presentations given at the hearings before the National Academies Hubble panel in 2004. Frank Cepollina and his office gave me a detailed chronology of their robot servicing mission effort.

In the years after 1993 I was a witness to many of the press events described in this book. For material prior to that time, Lynn Barranger of the institute's News Video office provided me DVDs of a number of important events in Hubble's history, including the April 8 and April 9, 1990, press briefings, the May 20, 1990, press briefing, the May 20, 1990, first light event, the June 27, 1990, press briefing, and the December 18, 1993, second light.

In addition to the many interviews I conducted myself, I also used the interviews held at several archives, including the Space Telescope History Project and the Space Astronomy Oral History Project (both at the National Air and Space Museum Archives), the American Institute of Physics Archives at the Niels Bohr Library, Center for the History of Physics in College Park, Maryland, and NASA Oral History Project interviews found in the NASA History Office archives in Washington, DC.

PERSONAL INTERVIEWS

Neta Bahcall, 10/5/2006; Steven Beckwith, 5/26/2004, 3/16/2006; Ralph Bohlin, 1/4/2006; Geoffrey Burbidge, 2/27/2006; Chris Burrows, 5/25/2006; Frank Cepollina, 10/2/2006; Art Code, 1/9/2006; Ted Cox, 1/3/2006; Jim Crocker, 9/15/2006; Kris Davidson, 9/19/2006; Rodger Doxsey, 2/2/2006; Sandra Faber, 2/28/2006, 4/18/2006;

Holland Ford, 10/13/2006; Riccardo Giacconi, 2/23/2006; Jeff Hester, 6/14/2005, 8/24/2006, 9/18/2006; Jon Holtzman, 5/23/2006; Tod Lauer, 5/18/2006; David Leckrone, 10/23/2006; Mario Livio, 2/2/2006; Don Morton, 1/5/2006; Max Nathan, 11/6/2006; Jeremiah Ostriker, 12/22/2005; Robert O'Dell, 1/10/2006, 4/21/2006; Nancy Roman, 1/9/2006; Doreen Spitzer/Lutie Spitzer Saul interview, 10/5/2006; Peter Stockman, 2/15/2006; Domenick Tenerelli, 1/10/2006; Ray Villard, 8/23/2006; Ed Weiler, 9/13/2006, 10/17/2006; John Wood, 1/31/2006.

SPACE TELESCOPE HISTORY PROJECT INTERVIEWS

Marc Aucremanne, 11/21/1983; John Bahcall, 11/3/1983, 12/20/1983, 3/22/1984, 3/28/1984, 8/26/1992; Neta Bahcall, 3/27/1985; William Baum, 6/22/1986; Michael Belton, 10/11/1984, 12/12/1984; Robert Bless, 11/3/1983, 2/21/1984, 11/11/1986, 9/25/1991; Al Boggess, 4/20/1984; John Brandt, 5/15/1989; Robert Brown, 4/3/1984, 6/7/1984; Bert Bulkin, 1/11/1985; Margaret Burbidge, 11/29/1984; Frank Carr, 3/7/1984, 5/29/1986; Clark Chapman, 10/10/1984; Art Code, 2/21/1984, 2/22/1984; Edward Danielson, 9/27/1985; Art Davidsen, 1/29/1984; Mike Disney, 5/1/1984; James Downey, 1/18/1984; Rodger Doxsey, 7/22/1987, 6/19/1992; Frank Edmondson, 6/8/1984; James Elliot, 11/21/1984; Garvin Emanuel, 12/2/1985; William Fastie, 6/4/1986, 6/6/1986; George Field, 3/10/1986; Don Fordyce, 10/31/1983; Laurence Fredrick, 5/2/1986; Riccardo Giacconi, 1/25/1984; Ed Groth, 3/15/1984; Arun Guha, 7/26/1983, 8/1/1983; Richard Harms, 5/26/1987; Richard Henry, 11/29/1983; Noel Hinners, 10/17/1984, 11/16/1984, 12/4/1984; Don Hunten, 12/14/1984; Sam Keller, 2/11/1985; Ivan King, 5/2/1984; Arthur Lane, 10/9/1984, 11/10/1984; Barry Lasker, 12/8/1983; Robin Laurance, 5/26/1983; David Leckrone, 8/14/1984; Malcolm Longair, 6/14/1984; John Lowrance, 3/28/1985; Duccio Macchetto, 4/10/1984; Bruce McCandless, 1/8/1986, 1/9/1986; Jesse Mitchell, 12/10/1984; Max Nein, 6/5/1984; Memphis Norman, 9/6/1984; 9/12/1984; Robert O'Dell, 5/21/1985; James Odom, 2/26/1985, 1/7/1987; Jean Olivier, 1/18/1984, 2/27/1985, 1/8/1987; Charles Pellerin, 8/1/1983; Arthur Reetz,

6/27/1983, 9/21/1983; Jack Rehnberg, 11/1/1983; Evan Richards, 2/20/1984, 6/11/1992; James Rose, 7/14/1987; Jeffrey Rosendhal, 9/12/1984, 10/1/1984, 12/23/1985; Jane Russell, 12/18/1984; Ethan Schreier, 7/6/1987; Daniel Schroeder, 10/7/1985; Thomas Sherrill, 9/24/1985, 7/21/1986; Pete Simmons, 9/10/1984, 9/11/1984; Stanley Sobieski, 2/8/1985; Fred Speer, 2/26/1985; Lyman Spitzer, 10/27/1983, 3/7/1984; Pete Stockman, 2/15/1985; Ernst Stuhlinger, 4/2/1984, 4/5/1984; John Teem, 9/26/1984, 10/4/1984; Rodger Thompson, 6/5/1992; Hedrick van de Hulst, 5/27/1983; Ed Weiler, 10/20/1983, 2/11/1986, 3/17/1986, 3/9/1987; James Welch, 8/20/1984, 2/12/1987; James Westphal, 9/28/1985, 5/15/1987; Richard White, 4/6/1984; Ray Zedekar, 1/8/1986.

SPACE ASTRONOMY ORAL HISTORY PROJECT INTERVIEWS

Art Code, 9/30/1982; Robert Frosch, 7/10/1981, 7/23/1981, 8/19/1981, 9/15/1981, 10/6/1981; Leo Goldberg, 2/22/1983; Noel Hinners, 7/31/1981; Robert O'Dell, 4/12/1982, 4/14/1982; 12/15/1982; William Pickering, 12/14/1982; Nancy Roman, 1/28/1983, 2/3/1984; Ronald Schorn, 7/27/1983; Martin Schwarzschild, 6/18/1982, 4/20/1983; Lyman Spitzer, 6/17/1982; James van Allen, 2/18/1981, 6/12/1981, 6/18/1981, 6/22/1981, 7/15/1981, 7/16/1981, 7/28/1981, 8/6/1981; James Webb, 7/22/1983; James Westphal, 8/9/1982, 8/12/1982, 9/14/1982.

AMERICAN INSTITUTE OF PHYSICS INTERVIEWS

Sandra Faber, 7/31/2002; George Field, 7/14/1980, 7/15/1980; Leo Goldberg, 5/16–17/1978; Jesse Greenstein, 4/7/1977, 7/21/1977, 7/26/1977, 5/19/1978; Ivan King, 7/18/1977; William Morgan, 8/8/1978, 8/9/1978; John Naugle, 8/20/1980; Homer Newell, 7/17/1980, 10/20/1980; Nancy Roman, 8/19/1980; Martin Schwarzschild, 7/30/1975, 3/10/1977, 6/3/1977, 9/27/1977, 12/16/1977, 7/19/1979; Lyman Spitzer, 4/8/1977, 3/15/1978, 5/10/1978, 11/27/1991; Alfred Whitford, 7/17/1977.

NASA HISTORY OFFICE INTERVIEWS

Homer Newell, 4/21/1976, 5/25/1977.

JOHNSON SPACE CENTER ORAL HISTORY PROJECT

Randy Brinkley, 1/25/1998.

PUBLISHED SOURCES

Alexander, D. W. *Hubble Space Telescope Thermal Cycle Test Report for Large Solar Array Samples with BSFR Cells*. Huntsville, AL: NASA TM-108373, Marshall Space Flight Center, 1992.

Allen, Lew, ed. *The Hubble Space Telescope Optical Systems Failure Report*. Washington, DC: NASA TM-103443, 1990.

Augenstein, Bruno W. *Evolution of the U.S. Military Space Program, 1945–1960: Some Key Events in Study, Planning, and Program Development*. Santa Monica, CA: RAND Corporation, 1982.

Bahcall, John, ed. *The Decade of Discovery in Astronomy and Astrophysics*. Washington, DC: National Academy Press, 1991.

————, ed. *Report of the HST-JWST Transition Panel*. 2003. Available from NASA at http://www.nasa.gov/pdf/49151main_hst-jwst.pdf as of 10/13/2006.

Bahcall, John, et al. "Hubble Space Telescope Images of a Sample of 20 Nearby Luminous Quasars." *Astrophysical Journal* 479 (4/20/1997): 642–658.

Ball, Robert S. *Great Astronomers*. London: Sir Isaac Pitman & Sons, Ltd., 1920.

Ball Aerospace and Communications Group. *Corrective Optics Space Telescope Axial Replacement Information Packet*. Boulder, CO: Ball Corporation, 1993.

Barbree, Jay, and Martin Caidin. *A Journey through Time: Exploring the Universe with the Hubble Space Telescope*. New York: Penguin Books, 1995.

Barfield, Claude E. *Science Policy from Ford to Reagan: Change and Continuity*. Washington, DC: American Enterprise Institute for Public Policy Research, 1982.

Baxter, James Phinney. *Scientists against Time*. Cambridge, MA: MIT Press, 1946.

Bely, Pierre-Yves, and Christopher J. Burrows. *The Next Generation Space Telescope: Proceedings of a Workshop Jointly Sponsored by the National Aeronautics and Space Administration and the Space Telescope Science Institute and Held at the Space Telescope Science Institute, Baltimore, Maryland, September 13–15, 1989*. Baltimore, MD: Space Telescope Science Institute, 1989.

Benvenuti, P., and Ethan Schreier, eds. *Science with the Hubble Space Telescope: Proceedings of ST-ECF/STScI Workshop, Chia Laguna, Sardinia, Italy, June 29–July 7, 1992*. Garching, Germany: European Southern Observatory, 1992.

Beyond Einstein: From the Big Bang to Black Holes. Structure and Evolution of the Universe Roadmap. Washington, DC: NASA NP-2002-10-510-GSFC, 2003.

Bloom, J. S., et al. "The Host Galaxy of GRB 970508." *Astrophysical Journal* 507 (11/1/1998):L25–L28.

Bok, Bart. *A Study of the Eta Carinae Region*. Groningen, Germany: Hoitsema Brothers, 1932.

Brown, Robert A., and Holland C. Ford, eds. *Report of the Hubble Space Telescope Strategy Panel: A Strategy for Recovery: Results of a Special Study August–October 1990*. Baltimore, MD: Space Telescope Science Institute, 1991.

Bulletin of the American Astronomical Society, vol. 25, no. 4, 1251–1506, 183rd AAS Meeting Abstracts, 1993.

Bulletin of the American Astronomical Society, vol. 35, no. 5, 1075–1496, 203rd AAS Meeting Abstracts, 2003.

Bulletin of the American Astronomical Society, vol. 37, no. 4, 987–1578, 207th AAS Meeting Abstracts, 2005.

Burbidge, Geoffrey. "A Speculation Concerning the Evolutionary State of Eta Carinae." *Astrophysical Journal* 136/1 (7/1962):304–306.

Burbidge, Geoffrey, and A. Hewitt. *Telescopes for the 1980s*. Palo Alto, CA: Annual Reviews, Inc., 1981.

Burbidge, Margaret, "Synthesis of the Elements in Stars." *Review of Modern Physics* 29/4 (10/1957):547–650.

Burke, Bernard, ed. *Space Science in the Twenty-First Century: Imperatives for the Decades 1995 to 2015*. Washington, DC: National Academy Press, 1988.

Burnham, Robert, Jr. *Burnham's Celestial Handbook: An Observer's Guide to the Universe beyond the Solar System*, 3 vols. New York: Dover Publications, Inc., 1978.

Bush, Vannevar. *Science, the Endless Frontier: A Report to the President on a Program for Postwar Scientific Research*. Washington, DC: National Science Foundation, 1945; reprinted 1960.

Buttman, Gunther. *The Shadow of the Telescope: A Biography of John Herschel.* New York: Charles Scribner's Sons, 1970.

Capers, Robert. and Eric Lipton. "Hubble Error: Time, Money, and Millionths of an Inch." *Hartford Courant,* 3/31/1991–4/3/1991.

Chaisson, Eric J. *The Hubble Wars: Astrophysics Meets Astropolitics in the Two-Billion-Dollar Struggle over the Hubble Space Telescope.* New York: HarperCollins, 1994.

Christianson, Gale E. *Edwin Hubble: Mariner of the Nebulae.* New York: Farrar, Straus and Giroux, 1995.

Clerke, Agnes M. *The Herschels and Modern Astronomy.* New York: Macmillan & Co., 1895.

Collins, Martin J. *Cold War Laboratory: RAND, the Air Force, and the American State, 1945–1950.* Washington, DC: Smithsonian Institution Press, 2002.

Columbia Accident Investigation Board Report, vol. 1. Washington, DC: Columbia Accident Investigation Board, 2003.

Committee on the Assessment of Options for Extending the Life of the Hubble Space Telescope. *Letter Report, July 13, 2004.* Washington, DC: National Academies Press, 2004.

———. *Assessment of Options for Extending the Life of the Hubble Space Telescope, Final Report.* Washington, DC: National Academies Press, 2004.

Crocker, James H. "Engineering the COSTAR." *Optics and Photonics News* 4/11 (11/1993):22–26.

Crowe, Michael J. *A Calendar of the Correspondence of Sir John Herschel.* Cambridge: Cambridge University Press, 1998.

Davies, John K. *Satellite Astronomy: The Principles and Practice of Astronomy from Space.* Chichester, UK: Ellis Horwood Limited, 1988.

Deutsch, Armin J., and Wolfgang B. Klemperer. *Space Age Astronomy, an International Symposium Sponsored by Douglas Aircraft Company, Inc., August 7–9, 1961, at the California Institute of Technology.* New York: Academic Press, 1962.

Dick, Steven. *Cancellation of the HST Shuttle Servicing Mission 4: Decision and Aftermath.* Paper and PowerPoint presentation presented at the AAAS meeting, Washington, DC, 2/18/2005.

Dickson, Paul. *Sputnik, the Shock of the Century.* New York: Walker & Company, 2001.

Downs, Anthony. *Inside Bureaucracy.* Boston: Little, Brown and Company, 1967.

Doyle, Robert. *A Long-Range Program in Space Astronomy: Position Paper of the Astronomy Missions Board, July 1969.* Washington, DC: NASA SP-213, Scientific and Technical Information Division, Office of Technology Utilization, 1969.

Dressler, Alan. *Voyage to the Great Attractor: Exploring Intergalactic Space*. New York: Alfred A. Knopf, 1995.

————, ed. *HST and Beyond: Exploration and the Search for Origins: a Vision for Ultraviolet-Optical-Infrared Space Astronomy*. Washington, DC: Association of Universities for Research in Astronomy, 1996.

————, ed. *Future Research Direction and Visions for Astronomy*. Bellingham, WA: International Society for Optical Engineering, 2002.

Dunkle, Terry, "The Big Glass." *Discover*, 7/1989, 68–81. Also available in revised and more complete form at http://www.terrydunkle.com/glass.php as of 8/8/2006.

Evans, David, et al., eds. *Herschel at the Cape: Diaries and Correspondence of Sir John Herschel, 1834–1838*. Austin, TX: University of Texas, 1969.

Fazio, G. G., et al., eds. *Astronomy from Space: Proceedings of the Topical Meeting of the COSPAR Interdisciplinary Scientific Commission E of the COSPAR 25th Plenary Meeting held in Graz, Austria, June 25–July 7, 1984*. Oxford, UK: Pergamon Press, 1985.

Fischer, Daniel, and Hilmar Duerbeck. *Hubble: A New Window to the Universe*. New York: Springer-Verlag, 1996.

————. *Hubble Revisited: New Images from the Discovery Machine*. New York: Springer-Verlag, 1998.

Fishman, Gerald. "Gamma-Ray Bursts: Observational Overview." *In Compact Stars in Binaries, Proceedings of the 165th Symposium of the International Astronomical Union, Aug. 15–19, 1994*, edited by J. van Paradijs et al., pp. 467–476.

A Forecast of Space Technology, 1980–2000. Washington, DC: NASA SP-387, Scientific and Technical Information Office, 1976.

Foster, Carlton, et al. "The Solar Array-Induced Disturbance of the Hubble Space Telescope Pointing System." *Journal of Spacecraft and Rockets* 32 (7/1995):34–644.

Freeman, Marsha. *How We Got to the Moon: The Story of the German Space Pioneers*. Washington, DC: 21st Century Science Associates, 1993.

Gamow, George. *My World Line*. New York: Viking Press, 1970.

Gaviola, Enrique. "Eta Carinae: I. The Nebulosity." *Astrophysical Journal* 111 (3/1950):408–412.

Glanz, James. "Accelerating the Cosmos." *Astronomy* 10/1999:44–51.

Goodwin, Simon. *Hubble's Universe: A Portrait of our Cosmos*. New York: Penguin Books, 1996.

Goodwin, Simon, and John Gribbin. *Deep Space: New Pictures from the Hubble Space Telescope*. London: Constable and Company Limited, 1999.

Government Accountability Office. *Costs of Hubble Servicing Mission and Implementation of Safety Recommendations Not Yet Definitive*. Report to the Senate Subcommittee on VA/HUD-Independent Agencies, 11/19/2004.

Gratton, L., ed. *Proceedings of the International School of Physics "Enrico Fermi," Course XXVII, Star Evolution*. New York: Academic Press, 1963.

Greenstein, Jesse L., ed. *Astronomy and Astrophysics for the 1970s: Report of the Astronomy Survey Committee*, 2 vols. Washington, DC: National Academy of Sciences, 1972.

Hall, Donald N. B., ed. *The Space Telescope Observatory: Special Session of Commission 44, IAU 18th General Assembly, Patras, Greece, August 1982*. Washington, DC: NASA CP-2244, Scientific and Technical Information Branch, 1982.

Hanle, Paul A., and Von Del Chamberlain, eds. *Space Science Comes of Age: Perspectives in the History of the Space Sciences*. Washington, DC: National Air and Space Museum, Smithsonian Institution Press, 1981.

Hargittai, Magdolna, and Istvan Hargittai. *Candid Science IV: Conversations with Famous Physicists*. London: Imperial College Press, 2004.

Harrington, Philip S. *Star Ware: The Amateur Astronomer's Ultimate Guide to Choosing, Buying, and Using Telescopes and Accessories*, 3rd edition. New York: John Wiley & Sons, Inc., 2002.

Harwood, William. "After High-Stakes Drama, Crew Removes Old Camera." *Spaceflight Now*, May 14, 2009. Available at http://spaceflightnow.com/shuttle/sts125/090514fd4/index2.html as of 9/10/09.

————. "Astronauts Turn to Muscle Power to Free Stuck Bolt." *Spaceflight Now*, May 17, 2009. Available at http://spaceflightnow.com/shuttle/sts125/090517fd7/index4.html as of 9/10/09.

————. "Grunsfeld Hails Hubble as Final Spacewalk Ends." *Spaceflight Now*, May 18, 2009. Available at http://spaceflightnow.com/shuttle/sts125/090518fd8/index4.html as of 9/10/09.

Hearings before the Committee on Aeronautical and Space Sciences, Senate, 93rd Congress, 2nd session, February 2, 26, 28, 1974.

Hearings before the Committee on Science and Technology, House, 94th Congress, 2nd session, January 27, 1976.

Hearings before the Committee on Science and Technology, House, 95th Congress, 1st session, February 1, 1977.

Hearings before the Committee on Science, Space, and Technology, House, 101st Congress, 2nd session, July 13, 1990.

Hearings before the Committee on Science, Space, and Technology, House, 103rd Congress, 2nd session, March 8, 1994.

Hearings before the Subcommittee on Appropriations, House, 93rd Congress, 2nd session, March 26, 27, 28, 1974.

Hearings before the Subcommittee on Appropriations, House, 94th Congress, 1st session, March 4, 5, 6, 1975.

Hearings before a Subcommittee of the Committee on Appropriations, House, 94th Congress, 2nd session, February 18, 1976.

Hearings before a Subcommittee of the Committee on Appropriations, House, 95th Congress, 1st session, March 29, 1977.

Hearings before the Subcommittee of the Committee on Appropriations, Senate, 103rd Congress, 2nd session, February 8, 1994.

Hearings before the Subcommittee on Manned Space Flight of the Committee on Science and Astronautics, House, 93rd Congress, 2nd session, February 19, 20, 21, 26, 27; March 5, 6, 1974.

Hearings before the Subcommittee on Space Science and Applications of the Committee on Science and Technology, House, 94th Congress, 2nd session, January 28, 29; February 3–5, 13–17, 19, 1976.

Hearings before the Subcommittee on Space Science and Applications of the Committee on Science and Technology, House, 95th Congress, 1st session, February 2, 4–7, 9, 1977.

Hearings before the Subcommittee on Space Science and Applications of the Committee on Science and Technology, House, 97th Congress, 2nd session, May 24, 25, 1982.

Hearings before the Subcommittee on Space Science and Applications of the Committee on Science and Technology, House, 98th Congress, 1st session, June 14, 16, 1983.

Hearings before the Subcommittee on Space Science and Applications of the Committee on Science and Technology, House, 98th Congress, 2nd session, May 22, 24, 1984.

Hearings before the Subcommittee on Science, Technology, and Space of the Committee on Commerce, Science, and Transportation, Senate, 95th Congress, 1st session, February 25; March 1, 3, 7, 9, 17, 18, 1977.

Hearings before the Subcommittee on Science, Technology, and Space of the Committee on Commerce, Science, and Transportation, Senate, 101st Congress, 2nd session, June 29, 1990.

Hearings before the Task Force on Physical Resources of the Committee on the Budget, House, 94th Congress, 2nd session, February 25, 1976.

Hearing of the Veterans Affairs and House and Urban Development and Independent Agencies Subcommittee of the Appropriations Committee, Senate, 108th Congress, 2nd session, March 11, 2004.

Herschel, Caroline. *Memoir and Correspondence of Caroline Herschel.* New York: D. Appleton and Company, 1876.

Herschel, John. *A Preliminary Discourse on the Study of Natural Philosophy.* New York: Johnson Reprint Corporation, 1966.

Hester, Jeff, et al. "Hubble Space Telescope Imaging of Eta Carinae." *Astrophysical Journal* 102/2 (8/1991):654–657, 835.

Hetherington, Norriss S., ed. *The Edwin Hubble Papers: Previously Unpublished Manuscripts on the Extragalactic Nature of Spiral Nebulae.* Tucson, AZ: Pachart Publishing House, 1990.

Hinners, Noel W. *Announcement of Opportunity for Space Telescope.* Washington, DC: NASA, 1977.

House Reports 13061–5, 93rd Congress, 2nd session, Report 93–1139, June 21, 1974.

House Reports 13134–8, 94th Congress, 2nd session, Report 94–1220, June 8, 1976.

Hubble, Edwin. *The Realm of the Nebulae.* New Haven, CT, and London: Yale University Press, 1982.

Hubble European Space Agency Information Centre. *Hubble: 15 Years of Discovery.* Garching, Germany: European Space Agency, 2006.

Hutchings, J. B., and S. G. Neff. "Optical Imaging of QSOs with 0.5 arcsec Resolution." *Astronomical Journal* 104/1(7/1992):1–14.

Kellerman, K. I., and J. M. Moran. "The Development of High-Resolution Imaging in Radio Astronomy." *Annual Review of Astronomy and Astrophysics* 39 (9/2001):457–509.

King-Hele, D. G., ed. *John Herschel 1792–1871: A Bicentennial Commemoration, Proceedings of a Royal Society Meeting held on 13 May 1992.* London: Royal Society, 1992.

Kirshner, Robert P. *The Extravagant Universe.* Princeton, NJ: Princeton University Press, 2002.

Kuiper, Gerard P. *The Atmospheres of the Earth and Planets*, revised edition. Chicago: University of Chicago Press, 1952.

Lawton, A. T. *A Window in the Sky: Astronomy from beyond the Earth's Atmosphere.* New York: Pergamon Press, 1979.

Leidecker, Henning, and Walter Thomas. "Notes on the Reliability of the HST Gyros." Published as part of the *NASA Electronic Parts and Packaging Program* and available at http://nepp.nasa.gov/index_nasa.cfm/993/ as of 2/28/2007.

Ley, Willy, and Wernher von Braun. *The Exploration of Mars*. New York: Viking Press, 1956.

Livio, Mario. "Hubble's Top Ten." *Scientific American*, 7/2006, 42–49.

Livio, Mario, Keith Noll, and Massimo Stiavelli. *A Decade of Hubble Space Telescope Science: Proceedings of the Space Telescope Science Institute Symposium, held in Baltimore, Maryland, April 11–14, 2000*. Cambridge, UK: Cambridge University Press, 2003.

Logsdon, John M., ed. *Exploring the Unknown: Selected Documents in the History of the U.S. Civil Space Program*, vol. V: *Exploring the Cosmos*. Washington, DC: NASA SP-2001-4407 History Office, Office of Policy and Plans, 2001.

Longair, Malcolm. *Alice and the Space Telescope*. Baltimore, MD: Johns Hopkins University Press, 1989.

Longair, Malcolm, and J. W. Warner. *Scientific Research with the Space Telescope: International Astronomical Union Colloquium Number 54, held at the Institute for Advanced Study, Princeton, NJ, August 8–11, 1979*. Washington, DC: NASA CP-2111, 1979.

Marshall Space Flight Center. *Large Space Telescope Phase A Final Report*, vols. 1–5. Huntsville, AL: NASA TM-X-64726, 1972.

McCurdy, Howard E. *Inside NASA: High Technology and Organizational Change in the U.S. Space Program*. Baltimore: Johns Hopkins Press, 1992.

McElroy, John H. and E. Larry Heacock, eds. *Space Applications at the Crossroads, Proceedings of American Astronautical Society Conference held March 24–25, 1983 at the Goddard Space Flight Center, Greenbelt, Maryland*. San Diego: Univelt, Inc., 1983.

McKee, Christopher F., and Joseph H. Taylor, Jr., eds. *Astronomy and Astrophysics in the New Millennium*. Washington, DC: National Academy Press, 2001.

McRoberts, Joseph J. *Space Telescope*. Washington, DC: NASA EP-166, Division of Public Affairs, 1982.

Meinel, A. B. "Astronomical Observations from Space Vehicles." *Publications of Astronomical Society of the Pacific*, vol. 71, no. 422, 10/1959, 369–380.

Moore, Duncan T., ed. *Final Report: Hubble Independent Optical Review Panel*. Greenbelt, MD: Goddard Space Flight Center P-442-0078, 1991.

Morse, Jon A., J. Michael Shull, and Anne L. Kinney, eds. *Ultraviolet-Optical Space Astronomy Beyond HST: Proceedings of a Meeting Held at the Regal Harvest House Hotel, Boulder, Colorado, USA, August 5–7, 1998*. San Francisco: Astronomy Society of the Pacific, 1999.

Morton, Don. "Recollections of a Canadian Astronomer." *Journal of the Royal Astronautical Society Canada* 97/24 (1993):24–30.

————. "Lyman Spitzer: Astronomer, Physicist, Engineer, and Mountaineer." In *Spitzer Space Telescope: New Views of the Cosmos*, Astronomical Society of the Pacific Conference Series 357/1 (2006), ed. L. Armus and W. T. Reach, 1–6.

Naugle, John E. *First among Equals: The Selection of NASA Space Science Experiments*. Washington, DC: NASA SP-4215, Office of Management, Scientific and Technical Information Program, 1991.

Nelson, Buddy, ed. *Hubble Space Telescope Servicing Mission 3B Media Reference Guide*. Bethesda, MD: Lockheed Martin, 2002.

Newell, Homer E. *Beyond the Atmosphere: Early Years of Space Science*. Washington, DC: NASA SP-4211, Scientific and Technical Information Branch, 1980.

The Next Generation Space Telescope: Science Drivers and Technological Challengers, Proceedings, 34th Liège International Astrophysics Colloquium, Liège, Belgium, June 15–18, 1998. Noordwijk, Netherlands: ESA Publications Division, 1998.

O'Dell, C. Robert. *Aerobatics Today*. New York: St. Martin's Press, 1984.

————. *The Orion Nebula: Where Stars Are Born*. Cambridge, MA: Belknap Press of Harvard University Press, 2003.

Odewahn, S. C., et al. "The Host Galaxy of the Gamma-Ray Burst 971214." *Astrophysical Journal* 509(12/10/1998):L5–L8.

Office of Space Science and Applications (OSSA). *Establishment/Clarification of NASA/OSSA Policies/Guidelines for the Operations for the Space Telescope Science Institute*, dated July 3, 1984.

Optical Telescope Technology: A Workshop Held at Marshall Space Flight Center, Huntsville, Alabama, April 29–May 1, 1969. Washington, DC: NASA SP-233, Scientific and Technical Information Division, Office of Technology Utilization, 1970.

Origins 2003: Roadmap for the Office of Space Science Origins Theme. Pasadena, CA: Jet Propulsion Laboratory, 2003.

Outlook for Space: A Synopsis, Report to the NASA Administrator by the Outlook for Space Study Group. Washington, DC: NASA Scientific and Technical Information Office, 1976.

Outlook for Space: Report to the NASA Administrator by the Outlook for Space Study Group. Washington, DC: NASA SP-386, Scientific and Technical Information Office, 1976.

Paczynski, Bohdan. "Gamma-Ray Bursts." In *Relativistic Astrophysics and Particle Cosmology*, edited by Carl W. Akerlof and Mark A. Srednicki. *Annals of the New York Academy of Sciences*, vol. 688, 1993, 321–330.

Pecker, Jean-Claude. *Space Observatories*. New York: Springer-Verlag, 1970.

Perlmutter, Saul, et al. "Discovery of a Supernova Explosion at Half the Age of the Universe." *Nature* 391(1/1/1998):51–54.

Peterson, Carolyn Collins, and John C. Brandt. *Hubble Vision: Astronomy with the Hubble Space Telescope*. Cambridge, UK: Cambridge University Press, 1995.

Piszkiewicz, Dennis. *Wernher von Braun: The Man Who Sold the Moon*. Westport, CT: Praeger Publishers, 1998.

Poelaert, D. *Hubble Space Telescope Solar Array: A Thermally Induced Disturbance Torque*. Paris: European Space Agency STM-238, 1987.

RAND Corporation. *The Rand Corporation: the First Fifteen Years*. Santa Monica, CA: RAND Corporation, 1963.

Regis, Ed. *Who Got Einstein's Office? Eccentricity and Genius at the Institute for Advanced Study*. Reading, MA: Addison-Wesley Publishing Company, Inc., 1987.

Richardson, Richard. "Luna Observatory No. 1." *Astounding Science Fiction*, 2/1940, 113–123.

Rieke, George H. *The Last of the Great Observatories*. Tucson: University of Arizona Press, 2006.

Rogerson, John. "Project Stratoscope—Solar Photographs from 80,000 Feet." *Sky and Telescope*, 1/1958, 112–115.

Ruderman, M. "Theories of Gamma-Ray Bursts." *Annals New York Academy of Sciences* 262(1975):164–180.

Russell, Henry Norris. "Where Astronomers Go When They Die." *Scientific American*, 9/1933, 112–113.

Schroeder, Daniel J. *Astronomical Optics*. San Diego: Academic Press, Inc., 1987.

Science, "News of Science," 7/10/1959, 89.

Schuiling, Roelof. "STS-61: Mission Report." *Spaceflight*, 3/1994, 78–81.

————. "STS-82: New Instruments for the HST." *Spaceflight*, 6/1997, 204–209.

————. "STS-82: Three EVAs Fix Hubble." *Spaceflight*, 3/2000, 97–103.

————. "Columbia Mission Upgrades Hubble." *Spaceflight*, 6/2002, 34–241.

Sembach, Kenneth R., et al., eds. *Hubble's Science Legacy: Future Optical/Ultraviolet Astronomy From Space: Proceedings of a Workshop Held at University of*

Chicago, Chicago, Illinois, USA, April 2–5, 2002. San Francisco: Astronomical Society of the Pacific, 2003.

Senate Reports, Miscellaneous Reports on Public Bills, VI, 13057–6, 93rd Congress, 2nd session. Washington, DC: Government Printing Office, 1974.

Senate Reports, Miscellaneous Reports on Public Bills, V, 13168–5, 95th Congress, 1st session, January 4–December 15, 1977. Washington, DC: Government Printing Office, 1977.

Shapiro, Stuart L., and Saul A. Teukolsky, eds. *Highlights of Modern Astrophysics: Concepts and Controversies*. New York: John Wiley & Sons, Inc., 1986.

Sidgwick, J. B. *William Herschel, Explorer of the Heavens*. London: Faber and Faber Limited, 1953.

Sietzen, Frank, and Keith L. Cowing. *New Moon Rising: The Making of America's New Space Vision and the Remaking of NASA*. Ontario: Apogee Books, 2004.

Simmons, F. Pete, ed. *Large Space Telescope: a New Tool for Science: AIAA 12th Aerospace Sciences Meeting, January 30–February 1, 1974*. Washington, DC: Symposium Organizing Committee, 1974.

Sky and Telescope, "Project Stratoscope Obtains Finest Pictures of the Sun," 12/1959, 79–80.

Smith, Bruce L. R. *The RAND Corporation: Case Study of a Nonprofit Advisory Corporation*. Cambridge, MA: Harvard University Press, 1966.

————. *American Science Policy since World War II*. Washington, DC: Brookings Institution, 1990.

Smith, Nathan, et al. *SN 2006gy: Discovery of the Most Luminous Supernova Ever Recorded, Powered by the Death of an Extremely Massive Star like Eta Carinae. Astrophysical Journal*, in press. Available at http://xxx.lanl.gov/abs/astro-ph/0612617 as of 5/3/2007.

Smith, Robert W. *The Space Telescope: A Study of NASA, Science, Technology, and Politics*. Cambridge, UK: Cambridge University Press, 1993.

Smith, Robert W., and Joseph N. Tatarewicz. "Replacing a Technology: The Large Space Telescope and CCDs." *Proceedings of the IEEE* 73/7(7/1985):1221–1237.

Smithsonian World. *Where None Has Gone Before*. Washington, DC: Smithsonian Institute, 1985.

Sobieski, Stanley, and E. Joseph Wampler. *Advanced Electro-Optical Imaging Techniques: Proceedings of a Symposium Held at NASA Headquarters, Physics and Astronomy Programs, Office of Space Sciences, September 22, 1972*. Washington, DC: NASA Scientific and Technical Information Office, 1973.

Space Science Board. *A Review of Space Research: Space Science Summer Study, at the University of Iowa, Iowa City, Iowa, June 17–August 10, 1962.* Washington, DC: National Academy of Sciences–National Research Council, 1962.

Space Science Board. *Space Research: Directions for the Future, Woods Hole, Massachusetts, 1965, parts 1–3.* Washington, DC: National Academy of Sciences–National Research Council, 1965.

Space Science Board. *Priorities for Space Research, 1971–1980.* Washington, DC: National Academy of Sciences, 1971.

Space Science Board. *Opportunities and Choices in Space Science, 1974.* Washington, DC: National Academy of Sciences, 1975.

Space Science Research in the United States: A Technical Memorandum. Washington, DC: Office of Technology Assessment, 1982.

The Space Telescope: Papers Presented at the 21st Annual Meeting of the American Astronautical Society, Denver, Colorado, August 26–28, 1975. Washington, DC: NASA SP-392 Scientific and Technical Information Office, 1976.

Space Telescope Science Institute. *Reports on Key Projects for the Hubble Space Telescope.* Baltimore, MD: Space Telescope Science Institute, 1985.

————. *Hubble Space Telescope: Fine Guidance Sensors Instrument Handbook, Version 2.1.* Baltimore, MD: Space Telescope Science Institute, 1990.

————. *The GO and GTO Observing Programs, Version 1.0.* Baltimore, MD: Space Telescope Science Institute, 1990.

————. *Hubble Space Telescope: Optical Telescope Assembly Handbook, Version 1.0.* Baltimore, Maryland: Space Telescope Science Institute, 1990.

————. *Hubble Space Telescope: Wide Field and Planetary Camera Instrument Handbook, Version 2.1.* Baltimore, Maryland: Space Telescope Science Institute, 1990.

————. *The GO and GTO Observing Programs, Version 3.0.* Baltimore, Maryland: Space Telescope Science Institute, 1992.

Spitzer, Lyman, Jr., "Perturbations of a Satellite Orbit." *Journal of the British Interplanetary Society* 9/3 (5/1950):131–136.

————. "Interplanetary Travel between Satellite Orbits." *Journal of the American Rocket Society* 22/2(3–4/1952):92–96.

————. "Astrophysical Research with an Artificial Satellite." In *Earth Satellites as Research Vehicles, Proceedings of the Symposium Held April 18, 1956 at the Franklin Institute in Philadelphia.* Philadelphia: Journal of the Franklin Institute, 1956.

————. "The Stellarator." *Scientific American,* 10/1958, 32.

————. ed. *Scientific Uses of the Large Space Telescope*. Washington, DC: National Academy of Sciences, 1969.

Spitzer, Lyman, Jr., and Jeremiah P. Ostriker. *Dreams, Stars, and Electrons: Selected Writings of Lyman Spitzer, Jr.* Princeton, NJ: Princeton University Press, 1997.

Spitzer, Lyman, Jr., and Franklin Zabriskie. "Interstellar Research with a Spectroscopic Satellite." *Publications of the Astronomical Society of the Pacific* 71/422(10/1959):412–420.

Stuhlinger, Ernst, and Frederick I. Ordway III. *Wernher von Braun: Crusader for Space*, 2 vols. Malabar, FL: Krieger Publishing Company, 1994.

Subcommittee on Space Science and Applications, *Space Telescope Cost, Schedule, and Performance Review*. Washington, DC: U.S. Government Printing Office, 1984.

Thackeray, Andrew D. "Identifications in the Spectra of Eta Carinae and RR Telescopii." *Monthly Notices of the Royal Astronomical Society* 113(1953):211–236.

————. *The Observatory*, 76(1956):103–105.

————. *Astronomical Spectroscopy*. New York: Macmillan Company, 1961.

Thompson, Allyn J. *Making Your Own Telescope*, Cambridge, MA: Sky Publishing Corporation, 1947.

Trimble, Virginia. "Martin Schwarzschild." *Publications of the Society of the Pacific* 109/742(12/1997):1289–1297.

Tucker, Wallace, and Karen Tucker. *The Cosmic Inquirers: Modern Telescopes and Their Makers*. Cambridge, MA: Harvard University Press, 1986.

van Genderen, A. M., et al. "Characteristics and Interpretation of the Photometric Variability of Eta Carinae and Its Nebula." *Space Science Review* 39 (1984):317–373.

Vaughan, Diane. *The* Challenger *Launch Decision, Risky Technology, Culture, and Deviance at NASA*. Chicago: University of Chicago Press, 1996.

Ward, Bob. *Dr. Space: The Life of Wernher von Braun*. Annapolis, MD: Naval Institute Press, 2005.

Warner, Brian, and Nancy Warner. *Maclear and Herschel: Letters and Diaries at the Cape of Good Hope, 1834–1838*. Cape Town, South Africa: A. A. Balkema, 1984.

Warren-Smith, R. F., et al. "Optical Polarization Map of Eta Carinae and the Nature of Its Outburst." *Monthly Notes of the Royal Astronomical Society* 187 (1979):761–768.

Weiler, Ed, and Karen Jensen. "Redemption." *Air and Space*, 10–11/1996, 16–18.

Westphal, James, et al. "Absorption-Line Redshifts of Galaxies in Remote Clusters Obtained with a Sky-Subtraction Spectrograph Using an SIT Television Detector." *Astrophysical Journal* 197 (5/1/1975):L95–L98.

Whitford, A. E., ed. *Ground-Based Astronomy: A Ten-Year Program. A Report Prepared by the Panel on Astronomical Facilities for the Committee on Science and Public Policy of the National Academy of Sciences*. Washington, DC: National Academy of Sciences–National Research Council, 1964.

Whitney, Charles. *The Discovery of Our Galaxy*. New York: Alfred A. Knopf, 1971.

Wood, H. John. "Hubble Space Telescope: Mission Update." *Optics and Photonics News* 5/8 (8/1994):9–13.

Zimmerman, Robert. "Stellar Vision." *The Sciences* 34/2(3–4/1994):7–8, 48.

————. "The Shadow Boxer." *The Sciences* 36/1(1–2/1996):16–19.

————. "On a Clear Day You Can See Forever." *The Sciences* 38/2(3–4/1998):13–16.

————. "There She Blows." *The Sciences* 40/1(1–2/2000):25–29.

————. *The Chronological Encyclopedia of Discoveries in Space*. Phoenix, AZ: Oryx Press, 2000.

————. "More Light." *Invention and Technology* 18/2(Fall/2002):14–23.

INDEX

Note: Page numbers in **bold** indicate illustrations.

Also by Robert Zimmerman

Genesis: The Story of Apollo 8

The Chronological Encyclopedia of Discoveries in Space

Leaving Earth: Space Stations, Rival Superpowers, and the Quest for Interplanetary Travel